Optimizing Project Management

Optimizing Project Management

Te Wu

CRC Press
Taylor & Francis Group
Boca Raton London New York

CRC Press is an imprint of the
Taylor & Francis Group, an **informa** business
AN AUERBACH BOOK

CRC Press
Taylor & Francis Group
6000 Broken Sound Parkway NW, Suite 300
Boca Raton, FL 33487-2742

Printed on acid-free paper

International Standard Book Number-13: 978-0-367-42992-8 (Paperback)

Visit the Taylor & Francis Web site at
http://www.taylorandfrancis.com

and the CRC Press Web site at
http://www.crcpress.com

To my parents, thank you for always being there for me, from raising me to be an independent thinker and the endless support over the years. To my children for your small and big sacrifices, such as weekends of staying home because I was too busy, coping with dad's cooking, or even putting up with my temper. To my close friends from the college years who have helped to shape my sense of self by giving me the platform to learn and experiment about respect, collegiality, trust, and building lasting friendships. Most importantly, to my darling wife who has always stood by me and supported me when I am down and giving me the love and space to pursue my passions. I am truly blessed with all of you in my life.

Contents

List of Figures

List of Tables

List of Templates in Appendix A

Preface

My primary motivation for writing this book is to provide an optimized way of performing project management. Over the past 20 years, project management has expanded and evolved rapidly, and there are a number of perpetual big questions in project management such as

- Why should the organization apply project management?
- What is the value of project management in the broader context of an organization?
- Is project management as successful as some advocates suggested? On the flip side, is project management a waste of time and resources because of the many extensive and bureaucratic processes?
- Which project management approach should our project team adopt: predictive or adaptive, waterfall or rolling water, extreme programming or Scrum?
- How is the body of knowledge, practices, and processes organized in project management?
- What are the key concepts, tools, and techniques that project managers need to understand and perform?
- What are some common challenges confronting projects?
- How does one build a career in project management? What are the potential roles? What are the certifications? Is there a career map?

This book strives to provide an optimized view of project management by balancing and blending between competing methodologies (traditional versus Agile), lengthy methodologies and broad principles, processes and practices, and the need to understand versus the need to apply. Consequently, this book is written in a modular format with twenty-one relatively short chapters and three appendices providing additional value. As a further example, Appendix B includes a case study with multiple projects from a fictitious global organization. My plan is to develop more case studies, based on industries and professions, to demonstrate how project management can be tailored and applied to various scenarios. For more information, visit www.optimizepm.com.

Audience

This book is written for a wide audience who are either aspiring or practicing project management professionals. The book covers the core concepts, practices, and skills that are useful for professionals to develop new ideas, plan activities, implement projects, and conduct planning and controlling of schedule, budget, and scope along the journey. The text may be particularly useful for students to learn project management, project professionals to refresh their knowledge, and pursuers of project management certifications such as PMI's Project Management Professional (PMP) or PMI's Certified Associates in Project Management (CAPM). Compared with the current *PMBOK® Guide* – Sixth Edition, for example, this book covers all ten of PMI's knowledge areas plus some additional knowledge domains. Therefore, this book is well organized and structured as a companion guide to the latest project management standard.

Content

This book is organized in four parts and twenty-one chapters. Within each chapter, the book largely embraced a Socratic approach of asking poignant questions, designed to stir the reader's curiosity and critical thinking. It also provides four callouts of valuable content including definition, insights from the front line, good practices, and tools.

The modularity of the book is illustrated in the three appendices:

- Appendix A contains a list of commonly-used project management templates for both predictive and adaptive project management approaches.
- Appendix B contains an integrated case study based on a fictitious global company undertaking a number of projects and confronting various project challenges. The case maps closely to the chapters and sections, key concepts, and relevant tools of this book. The goal is to provide instructors, students, and practitioners with a realistic project example to practice and apply the key learnings in the book. For more information and additional case studies that the author plans to create over time, including potential collaboration opportunities, visit www.optimizepm.com.
- Appendix C contains a glossary of selective terms, reprinted with permission from Dr. Te Wu and Mr. Brian Williamson's book titled *The Sensible Guide to Key Terminologies in Project Management* (iExperi Press, Montclair, NJ).

www.optimizepm.com

To make project management applicable to a wide range of industries and professions, I have also created a website to support this book and the continuing evolution of the profession. There, readers can find project management templates, additional case studies, and other valuable information for practitioners, students, and educators.

I also envision that this book is just the first step in the attempt to optimize project management. As the profession becomes ever more integrated into the very fabric of businesses and organizations and with the rapid advancements in technology, the discipline of project management will change quickly, so this book can serve as a platform for collecting and sharing ideas to further enhance how we can all optimize project management for our projects.

Acknowledgments

First and foremost, I would like to thank my family – my dear wife and my children – for their support. Without them giving me words of encouragement and forgiving me for lapses big and small, I would not have found time to complete this book.

Next, it is vitally important to thank everyone who has provided inputs, insights, and even criticism of the many ideas contained in this book. This includes my students, colleagues, trainers in my firm, professionals in project management, and friends. There are too many of you to mention, but you know who you are.

Finally, I want to extend my thanks to Dr. John Wyzalek and everyone at CRC Press/Taylor & Francis Group for their effort and support. This includes providing editorial direction, guidance, and management of the book's development process. I am looking to build and strengthen our relationship.

Author

Dr. Prof. Te Wu is the founder and CEO of PMO Advisory, one of the most specialized project management consulting and training firms. On consulting, the firm concentrates on strategic business execution, specializing in PMO and project management. On training, the firm is a Project Management Institute's Registered Education Provider (PMI® R.E.P.) and is one of the first organizations to offer PMI certification training in Portfolio Management Professional (PfMP®), Program Management Professional (PgMP®), Project Management Professional (PMP®), Risk Management Professional (PMI-RMP®), and Agile Certified Practitioner (PMI-ACP®) training. In addition, the firm also offers training in organization change management, PMO, business management, and strategic business execution. The firm is a socially progressive firm balancing the goals of profitability with social aims. Dr. Wu is also a professor at multiple universities including Montclair State University (second largest university in New Jersey) and China Europe International Business School (CEIBS is ranked number 5 at the top MBA program in the world by the *Financial Times* in 2019). Previously, he also taught at Stevens Institute of Technology and Touro Graduate School of Business.

With more than 25 years of experience of helping businesses to improve their strategy execution, Dr. Wu is one of the few in the world with these professional certifications: portfolio management (PfMP), program management (PgMP), project management (PMP), and risk management (PMI-RMP). He is also an award-winning project manager, earning Honorable Mention in 2015 Project of the Year by the PMI-NJ Chapter.

Dr. Wu holds a doctorate degree with a dual concentration in management and international business from Pace University, two master degrees in industrial engineering and MBA from Columbia University and University of Phoenix, and two bachelor degrees in chemical engineering and philosophy from Stevens Institute of Technology. In addition to running PMO Advisory and teaching, he has served on Project Management Institute Global's Portfolio Management and Risk Management standard committees as a core member, and he is currently active on a number of task teams. He is an active board member of the New Century Education Foundation. In his spare time, he likes writing business articles and blogs, speaking at conferences especially when there are opportunities to expound on the value of project management, traveling to different corners of the world with his family, and just sitting back and reading good books and watching Netflix.

PMI, PfMP, PgMP, PMP, PMI-ACP, PMI-RMP, and PMI R.E.P. are registered marks of Project Management Institute Inc.

1

SETTING THE STAGE

Chapter 1

Project Management – What and Why?

Summary

This chapter provides an introduction to and overview of the field of project management. In addition to describing what is project management, explaining why it is important, and portraying the role of a project manager, this chapter also discusses the various stakeholders and responsibilities involved in managing projects, the career progression of project managers, and the skills, attitude, and behaviors that make an effective project manager. We will end the chapter with an overview of the Project Management Institute® (PMI). On completion of the chapter, you will better understand the why, what, who, where, when, and how of project management.

This chapter addresses these three sets of important questions:

1. Why! Why is project management important? Why is it worth your time pursuing a deeper understanding of project management? What value does it bring to organizations practicing project management?
2. What! What is a project? What is project management? What is project success? What is the PMI?
3. Who! Who is a project manager? What are their roles and responsibilities? Whom do project managers work for and with?

1.1 Importance of Project Management

1.1.1 Overview

Many organizations are not facing a shortage of ideas or even good ideas. Rather, organizations are struggling to turn their ideas into real, tangible results and to produce positive change that enables the viability and competitiveness of their organization. Project management is the discipline of choice that increases the likelihood of success and achieves positive outcomes.

A number of large-scale research initiatives have highlighted the need for project management and the value that this discipline holds in realizing results. In their 2017 Talent Gap Report,[1] the PMI®, one of the most internationally reputable organizations for project management, projected that there will be about 88 million individuals in project-oriented roles globally by 2027, an annual growth of 2.2 million roles every year from 2017 to 2027 with almost 214,000 in the United States alone. Schoper (2018)[2] discussed the projectification of the developed economies and estimated that project-related work is roughly 30% of the total economy.

One of the primary reasons that project management is gaining such recognition beyond niche operations is that organizational leaders are starting to recognize their key weaknesses in poor strategy execution. As many as 47% of leaders believe business execution is extremely important to their organizations,[3] and yet, 400 global CEOs have named execution excellence to be one of the greatest challenges they face[4]. This is probably because around nine out of every ten strategic implementations, which are mainly composed of projects, fail[5]. This is an astonishingly low success rate of just 10%.

The results of the 2015 Standish Chaos Report[6] point to a similar poor state of information technology project implementation. Over a period of 5 years, their research consistently found that fewer than 32% of information technology (IT) projects are rated as successful. The research also uncovered a relationship between the size of a project and the likelihood of its success. In general, the larger the project,

---

1 Project Management Institute, 2017. Talent gap: Job growth and talent gap 2017–2027, Project Management Institute, 2017. Retrieved from www.pmi.org/learning/careers/job-growth on November 20, 2019.

2 Schoper, Y.G., Wald, A., Ingason, H.T. & Fridgeirsson, T.V. 2018. Projectification in western economies: A comparative study of Germany, Norway and Iceland, *International Journal of Project Management*, 36(1), 71–82.

3 Strategic Business Execution Survey, August 15, 2014, PMO Advisory LLC.

4 Sull, D., Homkes, R. & Sull, C., 2015. Why strategy execution unravels—And what to do about it. *Harvard Business Review*, 93(3), 57–66.

5 Speculand, R., 2009. Six necessary mind shifts for implementing strategy. *Business Strategy Series*, 10(3), 167–172.

6 Standish Group 2015 Chaos Report—Q&A with Jennifer Lynch, October 4, 2015.

the more likely it was to fail; the smaller the project was, the greater the chance of its success.

Furthermore, there is a wide recognition of the applicability of project management to various industries and functions. Since ancient times, project management activities have been utilized in large construction projects, military campaigns, and public works. The construction industry is one of the first industries to embrace project management dating back to the dawn of civilization. Starting in the late 1960s, IT embraced project management for software development. As systems became more complex, project management became more popular fueling the growth of the PMI® and the Project Management Professional (PMP)® certification. Today, project management is commonly utilized in nearly all industries and functions.

1.1.2 Project Management and Its Environment

Why has project management been identified as a solution to the broadly expressed organizational challenge of strategic implementation and a vehicle to implement change? In fact, how does project management even relate to strategy?

Prior to the 1980s, many organizations sidelined projects as a tool for incremental enhancements and viewed project management as a part of operational management. Resources were primarily dedicated to operations and planning. But the relationship between projects, their management, and organizational execution is much stronger than many would realize, as projects became the dominant mechanism for implementing change. And if Heraclitus felt that "change was the only constant in life" in 535 BC, then he might be astounded at just how acute his observation has become in the 21st century. Today, in a globally connected marketplace, rapid technological advancements, increased complexity around sustainability, social turbulence, and regulatory impacts have created a perfectly dynamic and competitive environment. This has led organizations to shift their focus toward continuously changing or risk-facing demise. The result is a corporate version of Darwin's "survival of the fittest", where the fittest organizations are agile, strong, and prepared for any range of changes that lie ahead.

Organizations perform three types of activities: planning, operating, and changing. *Planning* focuses on what organizations should do, which can take on a number of forms, from annual budgeting to strategic planning. From the simplest of organizations, such as a corner grocery store where the shopkeeper counts the quantity of candies sold and in need of replenishment, to the most complex global enterprises in which large departments of people work on short-, medium-, and long-term objectives, planning is always a primary organizational activity. To maintain their viability, organizations must also operate their core functions, such as selling candies in the case of a grocery store, to global organizations manufacturing, distributing, and delivering complex products and services. *Operating* activities are focused around the generation of revenue and funneling profits to improve and renew the organization through change. *Changing* is the last of the three main functions. Activities for change are now a necessity. These activities are focused on creating new products and services, improving existing processes to be more effective and efficient, and striving to be ever more competitive through creating sustainable advantages. Projects are the vehicle to undertake unique and temporary endeavors, and even though projects can occur during all three activities (planning, operating, and changing), it is in the changing activities where most of the project intensity occurs.

As organizations have embraced change for competitive survival, they have been experiencing a surge in the number of projects implemented each year. Projects are no longer reserved for incremental enhancements. Rather, they have become the active mechanism to integrate new technologies, apply systems for sustainability, and mobilize knowledge sharing and many other complexities required for effective change. The successful management of these projects, and how they are implemented, makes the difference between a good idea and an effective outcome. This is where the role of the project manager becomes so valuable. See Table 1.1 for examples of projects.

Refining organizational objectives for change and growth, such as an expanding market share or restructuring for efficiency, is difficult. Possibly more difficult is developing a strategy for actually achieving objectives. But where the greatest challenge often lies is in practically executing on the projects that support this strategy. Skilled project managers are able to manage the daily microaspects of projects, lead teams on larger initiatives, and execute organizational strategies, thereby transforming strategies into tangible performance and meaningful outcomes.

Table 1.1 Example of Projects

Day-To-Day Project Management	*Managing Major Initiatives*	*Strategy Execution*
• Working on HR initiatives • Developing and executing a marketing plan • Managing initiatives in the supply chain • Coordinating legal work	• Information technology projects • Business process reengineering • Training and development • Project Management standard development • Complex enterprise resource planning	• Aligning business strategy with execution • Prioritizing initiative • Implementing portfolio management • Balancing risk with rewards

1.2 What Is Project Management?

In this book, a project is defined as "a time-limited, purpose-driven, and often unique endeavor intended to create an outcome, service, product, or deliverable."[7] Unpacking this definition reveals that a project has a defined start and end date (i.e., time-limited), one-of-a-kind requirements or context (i.e., unique), and delivering a particular outcome (i.e., purpose-driven). These three characteristics imply a certain level of complexity in managing projects within a set time frame, without precedent. The goal of project management is to apply relevant management tools and techniques for navigating this complexity and achieving project success through efficient and effective implementation. Project managers are the "scientists" exercising technical skills and tools, as well as the "artists" engaging in soft skills to manage stakeholder expectations, negotiate resources, channel team conflict and motivation, and tailor their approach according to the situation and its environment. The sponsoring organizations want to achieve specific deliverables, outcomes, benefits, or values.

Definition: A project is a time-limited, purpose-driven, and often unique endeavor intended to create an outcome, service, product, or deliverable.

1.2.1 Dimensions of Project Management

One of the most fundamental and traditional models in project management theory and practice is the Theory of Triple Constraints, often referred to as the Iron Triangle. This theory binds projects within three dimensions, namely, time, scope, and cost. For many years, project managers globally have worked within these dimensions to evaluate project progress and targets based on scheduling, resource consumption, and quality of output. The major assumption of this model is that change in one dimension will initiate a change in at least one of the other dimensions. For example, reducing the project budget can lead to a reduction of features (scope). The aim of the project manager is to achieve an optimal balance between these dimensions.

However, as the internal and external environment has become more complex and less consistent, organizations have come to require a more sophisticated, holistic view of projects. Today, organizations are looking beyond the Iron Triangle to additional dimensions, such as process maturity, stakeholder relationships, governance and decision-making, planning for and reacting to risk, and strategic alignment. A more complete spectrum of dimensions is shown in Figure 1.1.

7 Williamson, B. & Wu, T., 2019. *The Sensible Guide to Key Terminologies in Project Management*, iExperi Press, Montclair, NJ. Glossary.

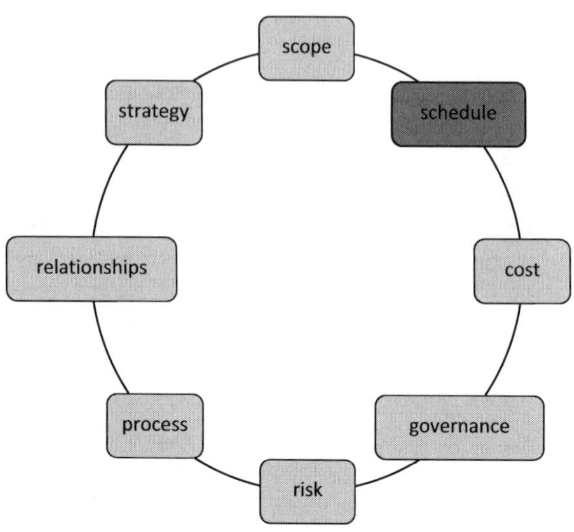

Figure 1.1 Project dimensions beyond scope, cost, and time.

On projects where these project dimensions have limitations, such as target completion date, set budget, or defined quality, these dimensions become constraints.

1.2.2 Recognizing a Successful Project

The Merriam-Webster Dictionary defines success as "favorable or desired outcome".[8] But "desire" is relative to the person who is "desiring". Even achievement can be perceived differently depending on the stakeholder. For example, while a chief financial officer might deem a project successful when it is completed within budget, the client might call the project unsuccessful because the product is of poor quality due to the limited resources that were allocated.

So how do project managers determine whether a project was successful or not? Many use the Iron Triangle to measure success. Those projects that come in on budget, on time, and aligned with scope requirements are often defined as successful. Sometimes, projects are believed to be successful when stakeholders feel generally positive about the process followed and the outcome achieved. And other times, projects are noted as successful, simply because they encountered no major operational issues. But the simplicity (and objectivity) of these success measures pose some questions.

- Is a project still successful if the project team is left burned out by the time of completion?
- What if a project is operationally faultless, but the deliverable fails to achieve its long-term goal?

[8] Merriam-Webster Dictionary's definition of success. Retrieved from www.merriam-webster.com/dictionary/success on October 30, 2019.

- How do you agree on the success of a project that was cost-effective, completed on time, and achieved the scope but failed to incorporate an added feature requested by the client?

In order to eliminate some of this ambiguity, there has been an increased focus on Total Project Success. This is a more holistic framework for measuring success that accounts for objective considerations such as schedule, cost, quality, scope, and resources, as well as subjective considerations such as team morale, client expectations, and a degree of integration. In Total Project Success, both project and project management are successful as shown in Table 1.2. This implies the successful delivery of key project deliverables and the suitable application of project management skills, processes, tools, and techniques to efficiently and effectively manage project implementation. Indications of a successful project can be product deliverables that are functioning as intended, customers who are satisfied with product performance, or the intended market embracing the product or service.

Insights from the front line. Project management success and project success are not the same thing. The world has witnessed many excellent projects delivered with arguably limited or poor project management.

Unfortunately, projects are often recognized more for their failure than for their success. While the potential reasons for project failure are vast and varied, some of the most common causes of failure are insufficient resources and time, unclear stakeholder expectations, frivolous scope changes during implementation, inadequate planning, and disruptive conflicts.

Project management is an organizational overhead, an overhead that often challenges the dominant organizational culture and policies in order to affect change.

Table 1.2 Project Success versus Project Management Success

Criteria	*Description*
Project or product success	• Meets needs • Deliverables used by customers • Satisfied customers • Improved market share • New technologies or capabilities
Project management success	• Robust planning • Rigorous examination of quality • Performance within cost and schedule • Well-executed project management process
Total success	• All the above

This can often lead to resistance from stakeholders and the user community in general. For this reason, project managers should tailor their application of project management processes to the project size, organizational culture, project type, and other unique considerations, such as the existence of project management offices (PMOs) in order to reduce unnecessary organizational complexity.

1.3 The Role of Project Manager

Projects can be large and complex and involve multiple stakeholders with varied expectations, roles, and interests that do not always align. The project manager is any person, sometimes informal, who is tasked with managing these stakeholders and interests such that the project achieves its desired outcomes. The project manager works with roles for oversight, such as sponsors and managers, as well as roles for implementation, namely the project team, in order to ensure that the end users are satisfied with the project deliverables (see Figure 1.2). In some projects, such as governmental public works initiatives, the end users can be the broader community who influence the project progress or benefit from the project deliverables. For example, when constructing a new highway, private homeowners are often asked to forfeit their land. But the larger society may benefit from the reduced traffic or a quicker travel route.

In order to manage these multiple and varied project management roles, project managers require a unique balance of hard and soft skills, listed in Figure 1.3. Technically, project managers should be able to analyze scope and create project Work Breakdown Structures, schedule resources, estimate and manage budgets, analyze risks and appropriate responses, and manage quality, including analyzing project components. On the softer side, project managers need to be able to guide, motivate, and monitor people at all levels: team members, vendors, and even senior managers and executives. To do this, project managers need to be equipped with skills in negotiation, decision-making, problem-solving, conflict resolution, and communication. These soft skills are especially important when "managing up". Engaging with project sponsors, executives, and even the PMO around project requirements and timelines is best handled carefully and convincingly.

1.4 About the Project Management Institute

The PMI® is the world's largest project management professional organization. They currently hold over 591,000 members across 215 countries and 309 chapters and potential chapters.[9] Their most popular and widely consulted publication is

® "Project Management Institute" and "PMBOK" are trademarks of the Project Management Institute, Inc.

[9] PMI Fact File, February 2020. PMI Today, Project Management Institute.

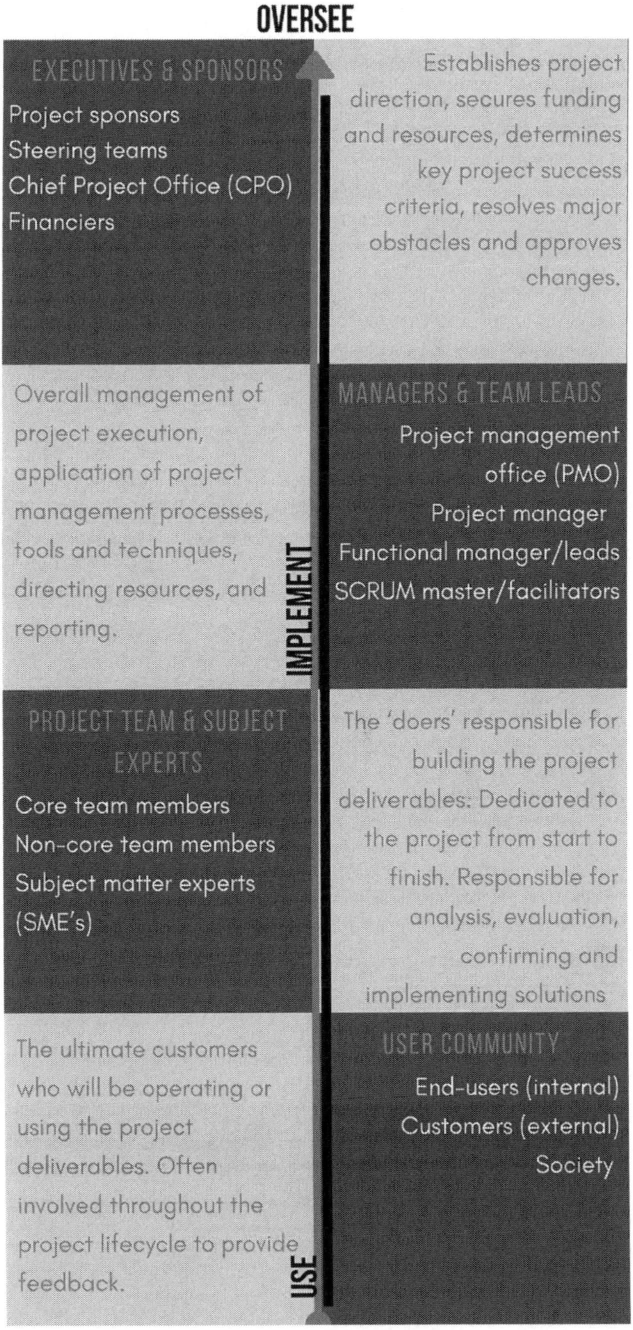

OVERSEE
EXECUTIVES & SPONSORS
Project sponsors
Steering teams
Chief Project Office (CPO)
Financiers
Establishes project direction, secures funding and resources, determines key project success criteria, resolves major obstacles and approves changes.
Overall management of project execution, application of project management processes, tools and techniques, directing resources, and reporting.
IMPLEMENT
MANAGERS & TEAM LEADS
Project management office (PMO)
Project manager
Functional manager/leads
SCRUM master/facilitators
PROJECT TEAM & SUBJECT EXPERTS
Core team members
Non-core team members
Subject matter experts (SME's)
The 'doers' responsible for building the project deliverables. Dedicated to the project from start to finish. Responsible for analysis, evaluation, confirming and implementing solutions
The ultimate customers who will be operating or using the project deliverables. Often involved throughout the project lifecycle to provide feedback.
USE
USER COMMUNITY
End-users (internal)
Customers (external)
Society

Figure 1.2 Project management roles. (Training and consulting content from PMO Advisory LLC. Reprinted with permission.)

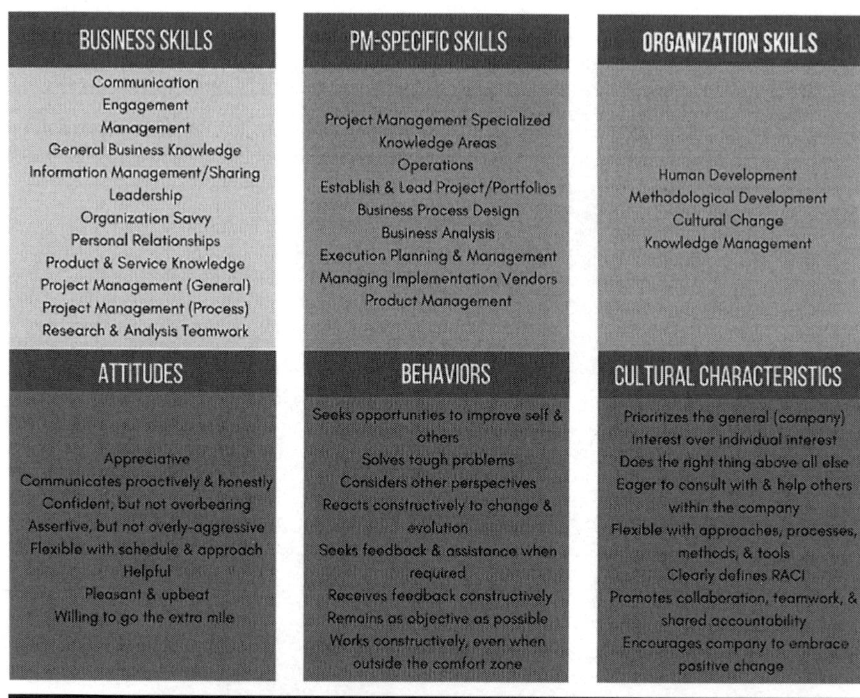

Figure 1.3 PM skills, attitudes, and behaviors. (Training and consulting content from PMO Advisory LLC. Reprinted with permission.)

A Guide to the Project Management Body of Knowledge (*PMBOK® Guide*), which is available in 12 different languages and has over 5 million copies in circulation today. Project management has been practiced for centuries (consider the construction of the ancient aqueduct system in Rome), but without a point of consistent research that leads to best practice. The *PMBOK® Guide* has provided a consolidated body of knowledge and best practice for the discipline of project management. While there are many guides and methodologies, such as PRINCE2*, the *PMBOK® Guide* remains one of the most internationally consulted standards.

Apart from the *PMBOK® Guide*, PMI* also provides opportunities for formal certification in the discipline of project management. Their certificates span vertically across levels of specialization and horizontally across areas of specialization. The most popular of the PMI certificates is the PMP®. In the past, this certificate was a means of gaining recognition and competitive advantage as a skilled project manager. However, holding the PMP credential today is a necessity for competitive parity in some sectors such as the United States Federal Government, rather than

* "PRINCE2" is a (registered) Trade Mark of AXELOS Limited. All rights reserved.

* "PMI", "PMP", "PgMP", "PfMP", "PMI-RMP", PMI-SP", "PMI-PBA", "PMI-ACP" and "CAPM" are service marks of the Project Management Institute, Inc.

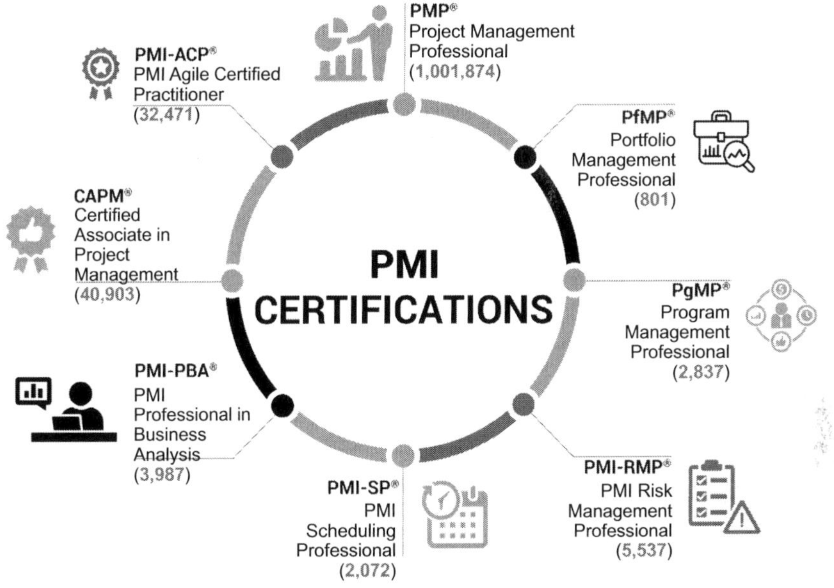

Figure 1.4 Active PMI certification holders. (January 2020). (PMI Fact File, February 2020. PMI Today, Project Management Institute.)

competitive advantage. Today, there are already thousands of people holding PMI® credentials, and this number is expected to grow significantly over the foreseeable years. Figure 1.4 shows a recent counting of active certification holders across the eight PMI certifications.

Attaining a PMI® certificate usually requires completing an application with prerequisite knowledge and/or related work experiences and passing an exam of between 120 and 200 competency-based questions depending on the certificate. Upon completing the application that includes requiring applicants to have completed a specified number of hours experience in project, program or portfolio management, and/or some level of formal training, the application can then take the exam. After successfully completing the exam, the applicant becomes an active certificate holder typically for a duration of 3 years after which the application must apply for recertification. The only current exception is the Certified Associate in Project Management (CAPM), which is a nonrenewable 5-year term.

1.5 Duty of Care

Project managers often confront difficult choices. For example, how often do you face the conundrum of "getting things done" versus "doing things right"; "putting out fires" versus "tackling the root causes"; or "toss projects over the fence" versus "thinking holistically". When reporting project status, should you report truthfully or should you "spin" even a little to avoid tough questions? If you find yourself plagued and even tortured by finding the best options when facing these difficulties and trade-off decisions, then congratulations and welcome to the project management profession. Studies have shown that while these can be painful choices, they are generally good and healthy signs that you care, because the alternate, apathy, is far more consequential and sometimes comes with disastrous results. As part of optimizing project management for total project success, it is important that project professionals care. Let's illustrate the concept of "duty of care" with two examples.

1.5.1 Case 1: Flint Michigan Water Crisis

The cover of Time Magazine on February 1, 2016, issue boldly titled, "*The Poisoning of an American City*", reminds us of the danger of apathy. This declaration brought clarity to the extent of failure at all levels of government and nongovernment organizations involved in one of the worst human created disaster in the United States in recent memory. How could something like this happen in the United States? As one read the story, the degree of ineptitude and the resulting failure at every level was astonishing and appalling. The full consequences of what went wrong and who were responsible are still unfolding after more than 3 years, and the investigation continues at the time of writing this book. Regardless of the many technical reasons, the predominant reason for this chain of systematic fiascos is a "lack of care".

Administrators, professional engineers, and probably some project managers involved in transitioning the water source from the Detroit water supply to the Flint River likely made some bad choices along the way – getting the work done so the city can start saving money – so flip the switch as quickly as possible. It is hard to imagine that no one involved questioned the quality of water from Flint River, but doing the easy thing won out. The water supply was switched, and the disaster unfolded.

The lack of successful prosecution at the time of writing showed that most government employees involved in this catastrophe had no malicious intent, at least at the onset. People on the Flint City Council, Dayne Walling (mayor at the time of the switch), emergency manager (Ed Kurtz at that time), and many others including the engineers and operators probably felt that they were "doing their best" to save Flint money. Likely, most of these people were acting with the best of intentions. But given this magnitude of the problem and the history of pollution of the Flint River, perhaps either the many concerned individuals maintained their silence or it was squashed by the upper management. When the problems were first reported, the government went through the classic five levels of incompetence:

1. **Deny.** The sample must be wrong. The sample size is too small. Did the sample come from water running after a few minutes (which of course reduces the amount of lead)?

2. **Cover up.** Michigan officials knew as early as the Summer of 2014 that something was wrong. Yet, the government stonewalled and withheld the information from public.
3. **Falsify.** As the situation got worse, officials started to change the process of water sampling and selective publishing data in favor of their position, further delaying actions.
4. **Deflect blame.** When the situation finally broke out into the public view late in the Fall of 2015, nearly everyone in the various level of government accused somewhere and someone else. Fingers pointed all direction except at oneself.
5. **Accept.** Finally when the implications of lead poisoning were clear to the nation, the government finally acted and declared a state of emergency on January 5, 2016. It's too late for the 100,000 Flint residents and the many more visitors to that city.

What's worse, it is heart breaking reading the story especially when a simple solution of adding anticorrosives costing the city of only $80–$100 per day can greatly reduce the amount of lead in water. Yet, it was not done due to a variety of reasons such as bureaucracy. This is institutional apathy at the worst. The civil servants and others involved in this entire chain of activities worried more about their bureaucratic processes, politics, and perhaps cost savings. They forgot their ultimate responsibility, which is to provide safe drinking water to their citizens. Without courageous people such as Dr. Mona Hanna-Attisha, the pediatrician who raised the issue into the public view, the crisis may have gone on for much longer.[10]

1.5.2 Case 2: Boeing 737 Max Series

On October 29, 2018, Indonesia's Lion Air Flight 610 flying in a brand new Boeing 737 Max 8 jet crashed just 13 minutes after takeoff into the Java Sea killing all 189 people onboard. After the incident, suspicion immediately honed on a new system, the Maneuvering Characteristics Augmentation System (MCAS) that were newly installed on the plane model. The initial focus was a lack of training and documentation. Boeing's CEO, Dennis Muilenburg defended the plane and touted the safety of 737 family of jets, Flight Crew Operations Manual, training, and the flight worthiness of Boeing 737 Max 8. The plane remained flying.

In less than 6 months later, on March 10, 2019, Ethiopian Airlines Flight 302 hurled toward the ground at nearly 600 miles per hour and eviscerated near the town of Bishoftu, Ethiopia, 6 minutes after takeoff, killing all 157 people aboard. Initially, the Federal Aviation Administration (FAA) continued to support the airworthy of Boeing 737 Max 8 and resisted the pressure to ground the plane.

---

[10] Gross, T., 2018. "Pediatrician Who Exposed Flint Water Crisis Shares Her 'Story Of Resistance'", NPR. Retrieved from www.npr.org/sections/health-shots/2018/06/25/623126968/pediatrician-who-exposed-flint-water-crisis-shares-her-story-of-resistance on October 30, 2019.

However, after Civil Aviation Administration of China grounded the plane on March 11, 2019, and shortly followed by every major aviation authorities around the world, the FAA acted 2 days later to ground all Boeing 737 Max 8 for commercial flights. Being the last major FAA to act, FAA gambled away its authority. At the time of writing this book (November 2019), the grounding of Boeing 737 Max 8 continues. European aviation authorities have already announced that they will evaluate the airworthiness independent of FAA decisions. Combined, the two crashes killed 346 people, severely crippled Boeing's reputation and eroded trust, and threatened FAA's status as the global aviation leader.

As this investigation is ongoing with new revelations coming out daily, some chilling findings that remind us the importance of a duty of care are as follows:

1. Reports are coming out that top Boeing pilots were complaining of "egregious issues" with Boeing 737 Max in 2016, 2 years before the crash.[11]
2. Wall Street Journal reported that after the first crash, Lion Air Flight 610, FAA's internal statistical models projected a high likelihood of a similar emergency within roughly a year. Yet, FAA did not consider the drastic step of grounding the plane. It simply reminded the pilots how to respond.[12]
3. Investigation after Lion Air crash showed that Boeing intentionally minimized documentation to save cost of training. Thus, in the case of Flight 610, the pilots were not even aware of the MCAS that took over the plane. In the case of Ethiopian Air Flight 302, the pilots were aware of the MCAS, but by the time they were able to deactivate the system, it was too late. In the test of Boeing jet, pilots had only 40 seconds to correct the error.[13]
4. Recent findings showed that Boeing consistently leaned toward the side of cost and time saving, so Boeing 737 Max series can compete favorably with Airbus A320 Neo, its closest, and only real competitor for narrow body plane. For example, while the original MCAS relied on two angle of attack sensors, the 737 Max relied on just one sensor. The indicator for MCAS was an "optional" feature that cost more, even though MCAS was installed on all Boeing 737 Max.[14] Combined with the strive toward minimizing training, Boeing is being blamed for emphasizing on cost saving over flight safety.

---

[11] Gelles, D. & Kitroeff, N., 2019. "Boeing Pilot Complained of 'Egregious' Issue With 737 Max in 2016", *New York Times*. Retrieved from www.nytimes.com/2019/10/18/business/boeing-flight-simulator-text-message.html on November 10, 2019.

[12] Pasztor, A. & Tangel, A., 2019. "FAA's move to keep MAX flying relied on Boeing—between the two crashes, the agency struggled to catch up", *Wall Street Journal*, Eastern Edition. New York.

[13] Nicas, J., Glanz, J. & Gelles, D., 2019. "In test of Boeing Jet, pilots had 40 seconds to fix error", *New York Times*. Retrieved from www.nytimes.com/2019/03/25/business/boeing-simulation-error.html on November 10, 2019.

[14] Tabuchi, H. & Gelles, D., 2019. "Doomed Boeing jets lacked 2 safety features that company sold only as extras", *New York Times*. Retrieved from www.nytimes.com/2019/03/21/business/boeing-safety-features-charge.html on November 10, 2019.

At the time of writing, U.S. Department of Justice is leading a criminal investigation of Boeing to examine how 737 Max aircraft was developed and certified.

Why are these examples relevant to all project managers? Simple. Project managers have a "duty of care" to their stakeholders, especially to the customers and ultimate users. Managing scope, schedule, cost, and the many other factors and project management processes are important. But none is more important than caring about the project and who the deliverables and outcomes will serve. Sometimes, project managers can overly rely on the project management processes, tools, and templates. But project managers should question if they are servicing the ultimate purpose of delivering the benefits. So the next time you are confronting difficult choices and having sleepless nights, pat yourself on the back. Because if you did not care, you would sleep quite soundly through that good night.

1.6 Why Project Management Is an Exciting Field

With the number of project management–related certificates available today, project managers have an exciting career trajectory, as they gain increased experience, education, and credentials. Figure 1.5 shows that as project managers make the move from beginner to expert and attain specialist certificates, they also assume increased accountabilities for larger, more complex projects. Organizations value project managers' abilities to work across an organization confronting multiple people, challenges, and risks, and for this reason, new and exciting roles often present themselves to skilled project managers. In particular, project management professionals have been found to earn on average 25% more than those without professional certifications.[15] Due to the perceived value of project management professionals in today's fast-paced and changing environment, people in this profession benefit from increased job security, as they are more difficult to outsource or replace with automation.

One of the most exciting aspects of project management is that no day, no person, and no project is the same. Perhaps, it is for this reason that Fredrik Haren describes project managers as "the most creative pros in the world".[16] The discipline brings new relationships to nurture, new problems to solve, and new solutions to implement. And with every small or large problem solved or risk averted, there are opportunities to celebrate success, new connections, and self-development. In an age when the fears of obsolescence due to technological automation run rife, project management may be an oasis among the storm. While no profession will be

---

15 PMI Annual Salary Survey—10th edition, 2017. Project Management Institute. Retrieved from www.pmi.org/-/media/pmi/documents/public/pdf/learning/salary-survey-10th-edition.pdf on November 10, 2019.

16 Project Management Point of View. Retrieved from https://gcimmarrusti.wordpress.com/pm-quotes/ on November 10, 2019.

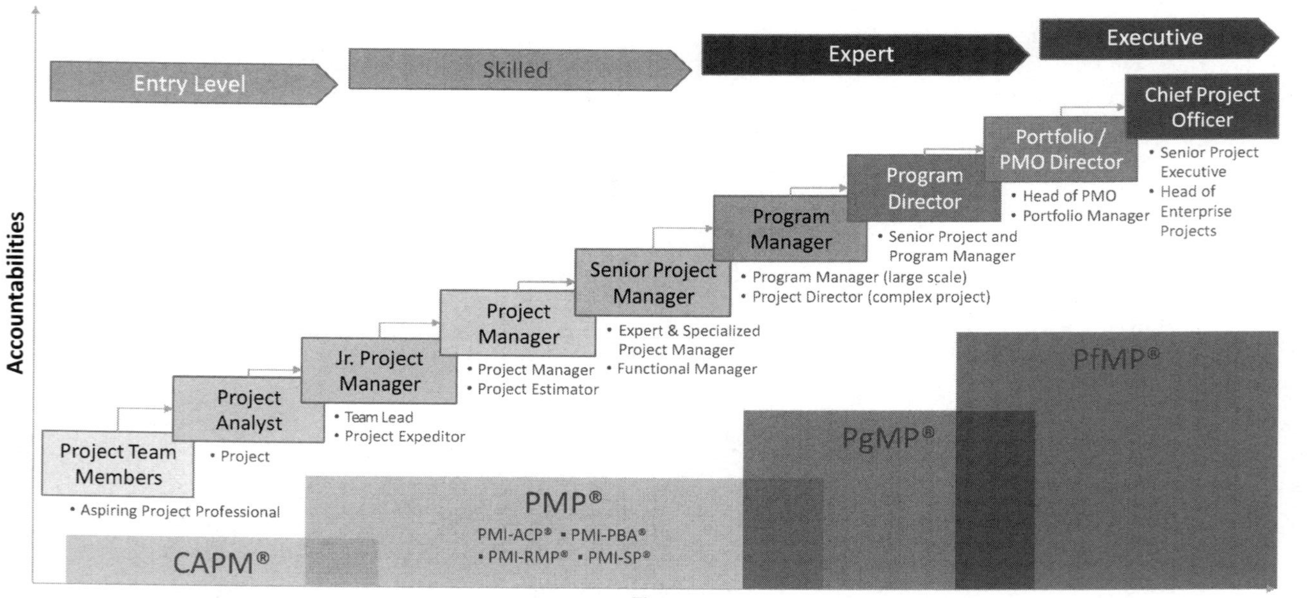

Figure 1.5 Project management career path. (Wu, T., 2017. *The Sensible Guide to a Career in Project Management*, iExperi Press, New Jersey. Reprinted with permission.)

insulated from technological advancements, project management's focus on people, problem-solving, and business judgment will be far more resilient than many other professions. Plus, automation is already greatly expanding the project manager's reach and efficiency. Furthermore, the field is advancing rapidly in the areas of Agile management; program, portfolio, and risk management; and also leveraging the Internet Web 3.0 technologies to improve collaboration and delivery,

Tool. If you are new to project management and wish to explore some projects that you have probably managed already, use Template 1: Identifying and Analyzing Projects on page 263.

1.7 Project Management in Motion

This book is intended for a wide range of audiences and across various industries and functions. Therefore, to balance between competing demands of ease of use, comprehensiveness, modularity, flexibility, and upgradeability, the author has adopted a modular format to this book. This includes the relatively short chapters on key topics. In addition, the appendices include three vital sections:

- Appendix A contains a list of commonly used project management templates for both predictive and adaptive project management approaches.
- Appendix B contains an integrated case study based on a fictitious global company undertaking a number of projects and confronting various project challenges. The case maps closely to the chapters and sections, key concepts, and relevant tools of this book. The goal is to provide instructors, students, and practitioners with a realistic project example to practice and apply the key learnings in the book. For more information and additional case studies in which the author plans to create over time, including potential collaboration opportunities, visit www.optimizepm.com.
- Appendix C contains a glossary of selective terms, reprinting with permission from Dr. Te Wu and Mr. Brian Williamson's book titled "*The Sensible Guide to Key Terminologies in Project Management*", iExperi Press, Montclair, NJ.

Chapter 2

Organizing Project Management Knowledge – Principles, Knowledge Domains, Life Cycles, and Agile versus Traditional Approaches

Summary

In the chapter, we discuss the organizing frameworks of project management. Understanding these frameworks is important to develop a deeper understanding of how the various knowledge domains, life cycle phases, processes, practices, and roles interact. These models will also demonstrate the evolving nature of project management both as new concepts are advanced and outdated thinking wanes.

There are many frameworks currently available, and some are more popular than others depending on countries, part of the world, industries, professions, and even companies. For example, companies such as Boeing, industry sectors such as construction, and professions such as cost accounting often develop their unique approach to project management. Furthermore, as project management itself is evolving, concepts from program management, portfolio management, and Agile approaches are also intertwining with project management.

This chapter strives to provide an optimal blend of multiple approaches including the following:

1. **Project Management Institute (PMI).** A Guide to the Project Management Body of Knowledge (*PMBOK® Guide*).
2. **International Standard Organization (ISO)** 21500:2012. Guidance on Project Management.
3. Combination of additional frameworks from program and portfolio management including concepts from the Strategic Business Execution framework.[1]

This chapter addresses these fundamental questions:

1. What are the important principles that guide project management?
2. How to organize the body of work in project management?
3. What are knowledge domains and project life cycle and why are they important?
4. What is Agile and how it differs from the traditional project management?
5. How to optimize project management through selecting the best project management approach and tailoring them to best fit your needs?

2.1 Project Management Principles

Project management is a specific branch of management formalized in the 1950s and is applied to initiating, preparing, implementing, transitioning, and/or closing projects, from the beginning to its natural conclusion. As a microcosm of management, project management focuses on the application of these management concepts often in a challenging environment constrained by a combination of purpose and scope, quality, schedule, resource, cost, risk, and environmental factors.

As a relatively new field, project management may appear to be unnecessary. After all, organizations today, especially prior to 1950s, can undertake all types of work, whether they are strategic or tactical, project or operational, purpose-driven or ad hoc without formal project management. But can they perform the work as competitively, efficiently, and effectively as with project management?

In an ever more cutthroat environment for not only businesses but also government agencies trying to do more with less, the successful organization is not only the one that delivers but also the one that delivers with certain advantages such

[1] Wu, T. & Chatzipanos, P., 2018. "Chapter 9: Achieving and Sustaining Execution Excellence Through Strategic Business Execution" in *Implementing Project Management: A Companion Guide to The Standard for Portfolio Management.* Project Management Institute, Newtown Square, PA. pp. 175–186.

as faster to market, lower cost, better quality, higher customer satisfaction, more functionality, or greater innovation. But project management has been expensive; organizations would need to invest in the development and training of project professionals, establish and utilize systems and tools, and develop and adopt processes. Not all onetime work should be managed as projects. Organizations and professionals adopting project management must weight both the advantages and disadvantages and tailor the application project management in their organization to achieve the optimized outcome.

Therefore, the challenge of project management is not only to deliver the project successfully but also to deliver the results at an optimal balance of cost, time, scope and quality, stakeholder satisfaction, and achieving short- and long-term objectives at an acceptable risk. To guide the optimization of project management, there are eight important principles to consider (Table 2.1).

Table 2.1 Key Project Management Principles

#	*Principle*	*Description*
1	Focus on performance and results	Projects are specific endeavors with one or more explicit purpose and desired outcomes. Project managers are the main drivers toward achieving those goals and objectives. Where possible, leverage lessons learned from past projects to enhance the likelihood of achieving results.
2	Minimize surprises	Project professionals, especially project managers, should look ahead when possible and plan the execution of a project. Irrespective of the project management approach, project manager's primary responsibility is to navigate the situation, establish and manage the processes, and drive toward results. Even positive surprises, or opportunities, can be perceived as poor project management if they occur unexpectedly as the project cannot take more advantage of the opportunity.
3	Manage responsibly	The challenge of project management is to do more with less. By being as effectively and efficiently as possible, effective projects management should deliver results at lower costs and with greater satisfaction. Project managers should also find the optimized method for implementing projects, such as selecting between the spectrum of approaches from the predictive to the adaptative.

(*Continued*)

Table 2.1 (*Continued*) Key Project Management Principles

#	*Principle*	*Description*
4	Optimize approach	Project management is an optimization exercise, performing the art of possibility by dealing with competing priorities and needs with factors such as availability of time and resources. Effective project management strives to find the optimal balance between good planning with overthinking and make difficult trade-off decisions, risk taking, speedy execution with thoroughness, focusing on exceptions while maintaining solid control over execution progress and managing change versus adopting change, and trust but verify. Combined with Manage Responsibly, project managers need to adopt the optimal method that works best for the project and the sponsoring organization.
5	Empower people	This is especially important on large projects involving many people. Organizations should provide an environment in which individuals can thrive and encourage their project managers to foster trust and independence in which people can contribute.
6	Communicate effectively	Communication has always been an important contributor of success. But in the era with a proliferation of technology tools such as social media, there may be a tendency toward too much information versus too little. Project professionals, especially the project managers, should concentrate on the most important messages and make sure they are delivered timely and appropriately.
7	Think and manage up	Most skilled project managers can manage both up and down one level effectively, but the truly experienced project managers can manage multiple levels up the organization chain. By developing the ability to think from the perspective of senior management, project managers can link their immediate project goals and deliverables with the broad goals and key performance indicators of executives multiple levels above their current standing. This is a sure way to ensure strategic alignment and continual support from the upper management.

(*Continued*)

Table 2.1 (*Continued*) Key Project Management Principles

#	*Principle*	*Description*
8	Fact-based management	Complex projects, especially projects in a politically tense environment, can be intricate to manage. Project managers should always focus on the facts first and consider them first in their decision-making processes. Other sentiments, can be important, and even if some ultimate decisions are political, project managers should be aware that their decisions are based on extrinsic factors beyond mere facts.

By following these eight principles, project professionals will increase their probability of successfully delivering projects with greater competence, efficiency, and effectiveness. Now with a good understanding of project management, the next section outlines the key knowledge domains.

2.2 Core Project Management Knowledge Domains

Project management knowledge domains are project management concepts, processes, and associated tools and techniques that have a specific project management application. These domains form the basic body of knowledge for project managers from the smaller and simple projects to the large and complex. Furthermore, these knowledge domains form the basis of "good" practices in project management, which will be elaborated in the future chapters. Table 2.2 describes the twelve common knowledge domains that are important for success management of projects, especially the larger and more complex projects:

Table 2.2 Common Project Management Knowledge Domains

#	*Knowledge Domain*	*Description*
1	Integration management	Enabling an integrated management of project management knowledge domains and processes across the life cycle
2	Stakeholders management	Engaging people and groups who can be influential on the project or impacted by the project
3	Scope management	Delivering the project purpose and specific requirements

(*Continued*)

Table 2.2 (*Continued*) Common Project Management Knowledge Domains

#	*Knowledge Domain*	*Description*
4	Schedule management	Managing time and sequence of tasks to ensure proper order of implementation
5	Resource management	Leading and directing people and other resources assets required for project execution
6	Cost management	Planning and monitoring budgets, project expenses, cash flow, and other financial considerations
7	Communication management	Interacting with stakeholders to ensure a common understanding and minimizing surprises
8	Risk management	Tackling uncertainties that may occur on projects, whether they are positive or negative events
9	Quality management	Examining the importance of quality including "fit for function" and "fit for use"
10	Supply chain management	Managing the entire chain of resources for the project. The emphasis is on directing the acquisition, purchase, and allocation of external resources such as contractors, equipment, and other assets
11	Conflict management	Confronting escalating disagreements, which is a specific form of issue, can occur on intense projects with severe constraints such as time, schedule, and resources. Disagreements over task priorities, clashes among people, and different preferences for the chosen project management approach can emerge
12	Governance management	Establishing and implementing effective decision processes to prioritize, analyze, and make decisions can be vital for project success. Depending on the organizational complexity and the political environment, instituting the optimal governance processes and structures can greatly impact the success execution of projects

Since projects are "artificial constructs created by humans", projects can different greatly even though they have the same goals. Projects can vary as greatly as the human imagination, and the above twelve knowledge domains are just the common ones. As project management evolves and as human creativity stretches in all directions, there can be many additional knowledge domains of consideration. These include other domains such as performance management, strategic alignment, capability and resource analysis, and benefits management. Although not infinite, the list can go on. But for this book, the focus is on these top twelve common knowledge domains.

Note: For those who plan to prepare for the PMI CAPM and PMP certification exams, it is important to understand the latest PMBOK® Guide fully. The current version of the guide contains 5 process groups (initiating, planning, executing, monitoring and controlling, and closing), 10 knowledge domains (the first ten knowledge domains specified above), and 49 processes.

2.3 Additional Knowledge Domains

Even though this book focuses on the top twelve knowledge domains that are the most common, it is important to note that there are additional knowledge domains that can and should be considered as a part of project management. Domains such as issue management, adoption management, or operation management are important areas for project managers to learn, as there are important questions and considerations that can influence project implementation.

#	*Knowledge Domain*	*Description*
13	Issue management	Addressing problems or obstacles quickly and effectively can minimize disruption to projects. Projects rarely, if ever, execute in an ideal state. Often, plans do not survive the first major challenge without causing change. Negative risks, such as threats, while they will ideally be successfully mitigated, can manifest and become issues.
14	Adoption management	Facing organizational and people challenges early on provides more time and space for project managers to maneuver and create adoption plans. Studies have shown that most projects ultimately fail not because of technology but because of people. Project managers should proactively plan adoption upfront; this can reduce project inefficiency and possible failure.
15	Operation management	Considering the project from a total life cycle perspective often includes the transitioning of project deliverables to operation. Project managers sometimes earn a poor reputation of focusing on the project implementation and neglecting what comes afterward. Most projects have significant postproject work, and the project team should be aware of this and plan it accordingly in the transition planning.

2.4 Project Life Cycle

Since projects are defined as time-limited endeavors, they must have both a beginning and an end. Well-managed projects will carefully progress through a number of phases. A representative and generic project, program, and portfolio life cycle can include these five phases: ideation, initiation, preparation, implementation, and transition and/or closure. Project managers should be able to tailor these phases and its intent to the specific requirements of their project or as dictated by the organizational processes.

For example, the life cycle of a project for creating training materials might include the following phases: (1) assessment, (2) content design, (3) creating materials, (4) piloting and training, and (5) finalizing the course. Or a software development project might follow the project life cycle phases: (1) define requirements, (2) design an architecture, (3) development, (4) quality assurance, and (5) deployment.

As project managers assess projects, identify important deliverables, and define activities required for project implementation, project managers should also determine the optimal number of project phases. Some considerations include the following:

- **Familiarity with the project.** For example, generic life cycle phases are ideal for unfamiliar projects that require progressive clarification or in organizations with strong project management offices to standardize the phases. Otherwise, for projects such as creating training materials, a more specific phase such as the one in the preceding paragraph is more relevant and easier to understand.
- **Logical grouping of project milestones and activities.** In some projects, the milestones and key activities can form the life cycle, such as the case with software development.
- **Ease of communicating with project stakeholders.** Project managers should design the optimal number of phases, often within a range of five plus/minus two (5±2), or three to seven (3–7), phases. Too many phases can be confusing and difficult to communicate. However, too few phases, such as starting and ending, often fail to capture the project complexity.

The traditional project management approach, also known as the predictive method, is based on a logical and progressive understanding of projects. The most common of the traditional approach is the waterfall method. The sequential phases outlined in Figure 2.1 show that proper planning of each phases and its project management deliverables will enable smoother downstream implementation. The disadvantage of this approach to project life cycles is its rigidity. This is especially problematic in a highly turbulent environment in which the requirements often change, or when customers cannot easily define the requirements in the first place.

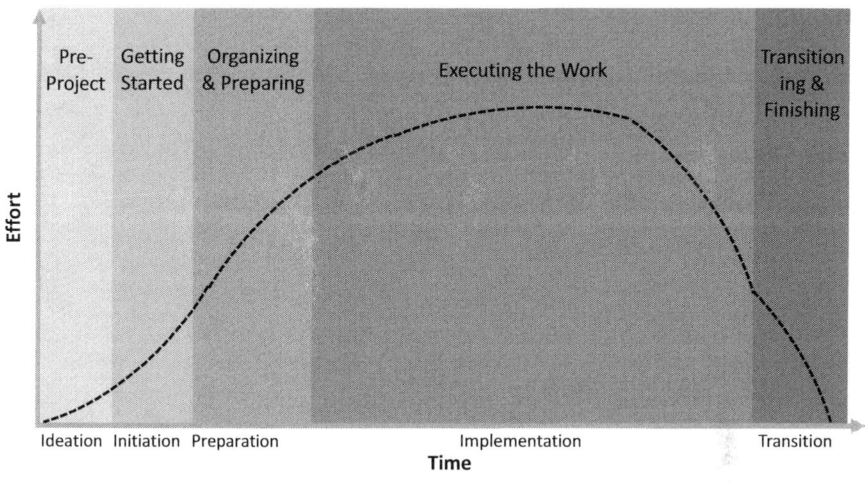

Figure 2.1 Traditional waterfall project life cycle. (Training and consulting content from PMO Advisory LLC. Reprinted with permission.)

Furthermore, as the traditional waterfall approach does not deliver project results until the entire project is completed, which can be multiple years, project failure is especially painful because of the investments made. Plus, in these long duration projects, both internal conditions and external environments can change, rendering the original requirements outdated in some mild cases or outright obsolete in some extreme situations. Therefore, too many executives, traditional waterfall projects can be inflexible and high risk.

In professions such as software development, professionals have adopted a more flexible approach where changes are more tolerated and, in some cases, even welcomed. Unlike the traditional waterfall approach, Agile approaches are more iterative and incremental, allowing for increased flexibility and adaptability to changing project requirements.[2] Figure 2.2 shows five iterations as "sprints" within the implementing phase of the project life cycle. Even if an iteration fails, the failure is smaller and improvements can be applied in the next iteration. By permitting and even encouraging the concept of "fail early, fail fast", studies have demonstrated that an Agile approach can improve the overall project results and decrease the risk of catastrophic failures.

[2] Griffin, C. & Roldan, M., 2013. *Swimming Up the Waterfall: Agile Processes in a Waterfall World.* Paper presented at PMI® Global Congress 2013—North America, New Orleans, LA. Project Management Institute, Newtown Square, PA.

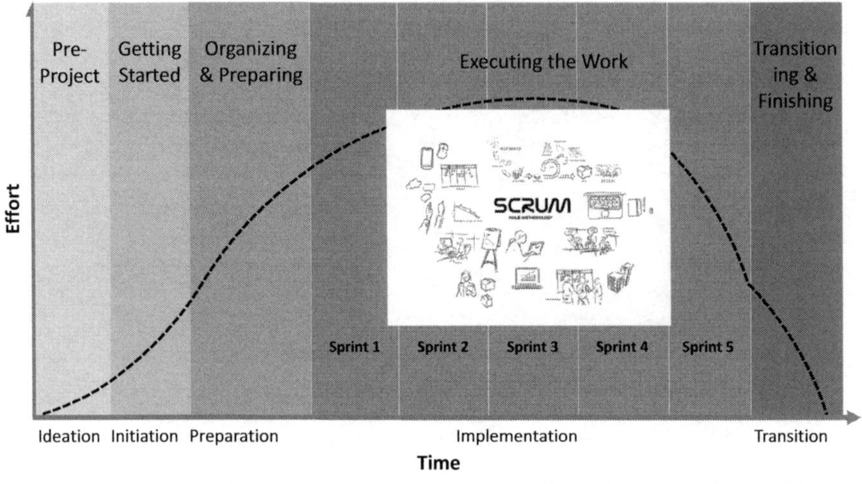

Figure 2.2 Project life cycle in an Agile approach. (Training and consulting content from PMO Advisory LLC. Reprinted with permission.)

2.5 Predictive Approach versus Adaptive Approach

Arguably two of the most popular approaches to project life cycles and planning are predictive and adaptive. A popular predictive project management approach is the waterfall project methodology. This method is one of the most commonly applied and traditional ways of thinking about project life cycles. Through progressive elaboration and preexecution analysis, waterfall attempts to understand and plan as many of the project deliverables upfront as possible. It encourages a thoughtful, detailed, and elaborate planning. Change is largely unaccounted for or unknown until it occurs; the waterfall approach prioritizes stability over the dynamism. But planning for change is an important aspect of traditional waterfall approach, as there can be unpredictability and changes during project implementation. This rigidity has posed obstacles for projects implemented in the uncertain and rapidly changing and advancing environment of contemporary projects. The thorough planning of detailed specifications is difficult at the projects' beginning. Even when performed correctly and thoroughly, requirements and scope are largely interpretive. For larger projects with long time horizons, the turbulence in the market often renders earlier requirements outdated, if not obsolete. It can also be difficult to obtain sign-offs from customers and sponsors so well in advance. In many cases, organizations spend years building systems that fail and waste precious resources. These frustrations have been particularly rife in the software development industry, where waterfall has been found to be too slow, disconnected, document-heavy, bureaucratic, and insufficient for responding to change.

Insights from the front line. In 2001, a group of software developers attempted to find a new, more flexible approach to projects. What they created was the Agile Manifesto, initially developed specifically for software development projects. The emphasis of Agile is on individuals and interactions over processes and tools, time spent on development over comprehensive documentation, and customer collaboration over contract negotiation, responding to change over following a plan.[3]

The promise of Agile is that change is a constant. By incorporating change as a parallel process component, Agile approaches to projects are iterative and incremental, rather than sequential. This provides three obvious benefits:

1. Even when projects fail, the failure is often smaller, affecting only the last iteration rather than the entire project.
2. Change is embraced as a normal part of project evolution and can thus be accounted for in the future iterations.
3. Value is also delivered incrementally, which allows sponsors to see value in days or weeks, rather than months or even years.

While adaptive methods such as Agile project management are growing in popularity, the approach is not easy to implement for many organizations. Misunderstanding of the approach often leads to poor project planning, documentation, and process rigor. This is largely because Agile requires fundamental changes in organizational culture and decision-making in order to implement a speedy and adaptable project life cycle effectively. For example, traditional, bureaucratic cultures where difficult decisions require months for sign-off will impede an Agile approach. Unless the project manager is a strong Agile practitioner and supported by an Agile project team and sponsorship, Agile remains more of an idealized principle than a practical method.

2.5.1 Derivative Methodologies

Waterfall and Agile are not the only project management methodologies used today. As the presence of change forces organizations to become more sophisticated, a range of methodologies for project management has evolved to include a combination of characteristics from waterfall and Agile including hybridization of methods. These methodologies can be understood in terms of their inclination toward either traditional or Agile approaches. Many organizations choose a methodology depending on the type of project or program. A more traditional, waterfall methodology is usually

[3] Manifesto for Agile Software Development, 2001. Retrieved from https://agilemanifesto.org/ on November 10, 2019.

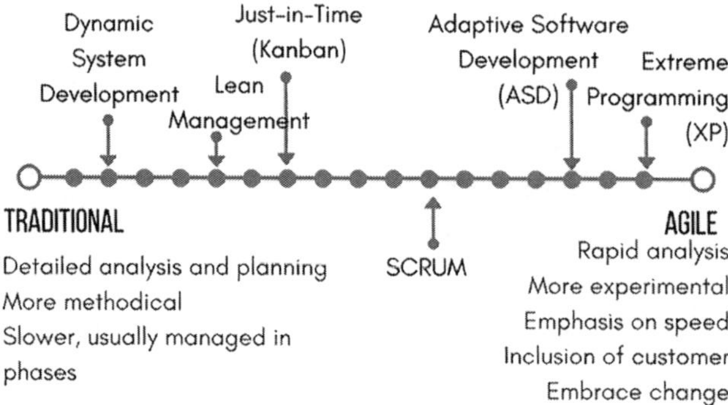

Figure 2.3 Continuum of traditional to Agile approaches. (Training and consulting content from PMO Advisory LLC. Reprinted with permission.)

preferred when the project requirements are clear and change is unlikely (major construction projects and specific software development). When change serves as the foundation for everything else (new software platforms or railways), and when definition, stability, and client expectations outweigh speed, particularly when an organization is still maturing, the Agile methodology is preferred. An Agile methodology is used when uncertainty is unavoidable due to ambiguous requirements or a dynamic and turbulent environment, when speed is more important than rigor, especially delivering value incrementally, when the project can be organized modularly, in iterations, or sprints, when organizations wish to adopt a "fail fast" approach, and when organizations possess the necessary maturity and culture for Agile. Many practitioners advocate a more blended or hybrid approach, striving to provide the optimal balance of stability with change. Figure 2.3 depicts a continuum of methodologies, ordered by their respective focus on traditional and Agile approaches.

Project professionals are regularly updating existing and developing new approaches and methods, and this is fueling the excitement of project management. The specific details of these methods are beyond the scope of the current book, and readers should visit www.optimizepm.com for the latest discussions and developments. This book strives to present a holistic model of how project management works with emphasis on the foundational concepts that strengthens the practice of project management.

2.6 Project Management Process

In addition to knowledge domains and life cycle, it is important to understand that project management contains a rational set of processes. A process is "a series of deliberate actions or steps taken in order to achieve a particular outcome or output".[4] Processes are important to project management, especially the longer-

[4] Williamson, B. & Wu, T., 2019. *The Sensible Guide to Key Terminologies in Project Management*, iExperi Press, Montclair, NJ. Glossary.

Table 2.3 Project Management Processes

Project Phase	*Process*
Ideation	Identify initial concerns and considerations pertaining to (knowledge domain)
Initiation	Identify and evaluate (knowledge domain)
Preparation	Develop plan for (knowledge domain)
Implementation	Execute plan, monitor progress, and control of (knowledge domain)
Transition and/or Closure	Perform closure or transition activities for (knowledge domain)

term success of organizational project management because establishing processes provide opportunities for improvements, to secure what worked well and to improve what should be enhanced. Organizations that consistently achieve superior project performance have built and improved the project management processes over time to achieve a higher degree of consistency and success, especially when efforts are made to improve continuously the project processes.

Various standards have proposed different processes. For example, in *PMBOK® Guide*, 6th Edition, there are 49 processes that sit at the interception of its knowledge domain and process group. This book creates a simpler framework for project management processes as shown in Table 2.3.

Depending on the project and the sponsoring organization, there may not necessarily be required processes for each of the knowledge domains. For example, on a small technical project that has no political and organizational implications, the set of governance management processes may be implicitly understood based on that organization's normal practices. However, for a merger and acquisition project that seeks to align titles and benefits of the merging organizations, governance management processes will probably be important to the project implementation.

2.7 Optimizing Project through "Right-Size" Project Management

Project management is not "one shoe that fits all". Organizations that inefficiently or ineffectively applied project management tend to rely on the "righteousness" of the processes and implement the full rigor of the processes dogmatically. This can overburden the project teams as well as the organization with bureaucratic processes. In some worst case, the process adds little value and so people reject them, thus missing out on the potential benefits. For example, should a small project of $20,000 with three people across 2 months and low impact have the same rigor of a business case or schedule planning as a major $2,000,000 project that span across a year with strategic implications? Or how many steps or documents required to start projects or report statuses in your organization? If the answer is "too many", then you are likely facing one or more of these dilemmas:

1. Balancing the need for processes but also delivering results. This is classically known as the "process vs. solution dilemma". In the current Project Management Institute's *A Guide to the Project Management Body of Knowledge* (*PMBOK® Guide*), there are nearly fifty processes, across ten knowledge areas and five process groups. It is important to remember that the processes themselves are only enablers to delivering results; processes are essential to efficiency and effectiveness of project operations. There are many project managers focusing too much on establishing and managing processes than balancing process with the need to roll up their sleeves and dive into the problems to solve them. This solution-oriented approach, especially on urgent and mission critical projects, is vital to a successful delivery, as miring too deeply into processes can make project management more bureaucratic than necessary.
2. Project manager's focus. What should be the focus of project managers – on managing all aspects of projects versus concentrating on the most vital few? Naturally, the answer is "it depends", and consider factors such as project uniqueness, process maturity, team composition, regulatory considerations, alignment of key project stakeholders, sponsorship, organization dynamics, vendors/contractors situations, project size and complexity, project management resources, and many more. The key is to analyze the situation and thoughtfully develop a healthy and balanced perspective. If you only have 8 hours in a day (or more like 16 hours for most project managers), what should you focus on?
3. Appropriate methodology. Today, there are a range of project management methodologies, from the adaptive such as Extreme Programming or Scrum to the predictive or traditional waterfall approach to managing projects. Combined with other project management disciplines such as program and portfolio management, project professionals have a range of choices including customizing the optimal approach for their projects. Selecting the inappropriate method can greatly burden the project team with inefficiencies. For example, on a website enhancement project in which the customer cannot make up their minds, using traditional waterfall approach will sack the team with slow change management and delivery. Worse, as the overall environment of websites is dynamic, by the time the website is built, it already became outdated. In these and similar projects where changes occur frequently, the Agile approach is likely to be more appropriate.

2.7.1 Four Guidelines for Right-Size Project Management

For project professionals seeking to find the optimal balance of efficiency and effectiveness, here are four guidelines to optimize project management for the right size:

1. Principle of proportionality
2. Focusing on the vital few
3. Find the best people
4. Execute flawlessly

The principle of proportionality is based on the fact that processes are both enablers and impediments, and it can be both at the same time depending on perspectives. In addition, as there are limited project management resources, chances are the more time that project managers work on process-related activities, the less time they have solving problems. No process or too much processes would likely result in confusion, inefficiencies, and in some cases conflicts. Most experienced project professionals have experienced situations in which they are forced to "follow the process" when the processes made little sense, both strategically and tactically. The magic is finding the balance, and it is recommended that project managers conduct a thorough due diligence of the project and determine the following:

1. What are the minimum success criteria for this project?
2. How important is the project to your client?
3. How much resources (e.g. budget, resources, and other scarce resources) is dedicated to this project?
4. What are the biggest risks?
5. What are the organizational climate and other environmental concerns?
6. How disciplined is the project sponsor, project team, and other important stakeholders?

Focusing on the vital few activities is essential especially when resources are finite. Working on too many activities simultaneously often means few or nothing are done well. While the theory is sensible, the practice of prioritizing is much more difficult in practice as what's high priority for one person may not be an equal priority for another. This is where strategy and project governance become crucial for organizations. Furthermore, the ability to focus on the vital few also enables organizations to adopt a more adaptive approach to managing projects. Thus, when organizations are able to prioritize and select few vital activities to implement, it can then build the iterative mechanisms to execute the next set of activities once the current set is done. This can greatly increase the organization ability of the project team.

When dealing with complex situations, whether it is in a highly competitive environment or when there are just a lot of projects, one of the best ways to mitigate (or exploit) risk is to recruit and nurture the best people. Managing large projects in multifaceted situations can be very challenging, and the unknowns are everywhere. The single best strategy is to hire the best people that the budget permits. Often, having more resources is not the most crucial, but having highly skilled, intelligent, and reliable professionals is essential. For Agile teams, this is especially important as cross-training is one of the most important ingredients for high-performance Agile teams.

The first three principles, starting from the macro to the micro, position the organization well on the road to achieve execution excellence. The last principle is all about executing the plan well by being reasonable with the resources and outline, assigning the best people, not only giving them space to explore but also providing the leadership to set direction, paying attention to details of the vital few but not losing sight of the complementary many, and trusting your strong people to execute but verify along the way.

Right-sizing project management is an approach to execution excellence. It is methodologically agonistic as it brings together "what works" from a variety of practices from traditional waterfall methodology to Agile and Lean management. It also places a lot of attention on people and organizations. The next time you work a project, ask yourself – is the size of project management processes suitable for your project? If not, then you can put these four guidelines to work.

2.8 Project Management in Motion

This book is intended for a wide range of audiences and across various industries and functions. Therefore, to balance between competing demands of ease of use, comprehensiveness, modularity, flexibility, and upgradeability, the author has adopted a modular format to this book. This includes the relatively short chapters on key topics. In addition, the appendices include three vital sections:

- Appendix A contains a list of commonly used project management templates for both predictive and adaptive project management approaches.
- Appendix B contains an integrated case study based on a fictitious global company undertaking a number of projects and confronting various project challenges. The case maps closely to the chapters and sections, key concepts, and relevant tools of this book. The goal is to provide instructors, students, and practitioners with a realistic project example to practice and apply the key learnings in the book. For more information and additional case studies in which the author plans to create over time, including potential collaboration opportunities, visit www.optimizepm.com.
- Appendix C contains a glossary of selective terms, reprinting with permission from Dr. Te Wu and Mr. Brian Williamson's book titled "*The Sensible Guide to Key Terminologies in Project Management*", iExperi Press, Montclair, NJ.

2

PROJECTS IN MOTION – FROM IDEAS TO RESULTS

Chapter 3

Ideation – Aligning Projects with Strategy

Summary

In this chapter, we discuss the first phase of project life cycle – Ideation. Ideation concentrates on aligning projects with the organization strategy to ensure superior execution. Furthermore, projects can also come from programs and portfolios supported by the management structures to enable strategic business execution. In an ideal world, there is a strong link between project selection and strategic planning, as well as the presence of good processes based on industry's best practices for selecting the most strategically relevant and feasible projects.

There are many tools for analyzing, filtering, and prioritizing ideas at every stage of evaluation. Common tools including the multicriteria scoring model, multicriteria weighted ranking, and single-criterion prioritization model are provided as examples. Finally, this chapter also discussed the importance of culture and how culture influences and shapes the optimal environment for project implementation.

Many organizations undertake projects that are questionable at best. These projects have a greater likelihood failure and waste organizational resources. It is therefore important to develop business cases to analyze ideas thoroughly before investing heavily in resources during project execution. This is the topic of the final section of this chapter.

This chapter addresses these fundamental questions:

1. What is strategic business execution and how are projects related?
2. Why do organizations choose some projects but not others?
3. What is ideation and why is it important?
4. What are the important components of a business case?

3.1 Strategic Business Execution and Environmental Context

Organizations are complex entities, and they perform a set of organizational activities to survive and thrive. These activities, as depicted in Figure 3.1, can be categorized into three dimensions of *planning*, *operating*, and *changing*.[1] Strategy is devised within the planning activities and executed in the form of projects in the changing and operating activities. Project, program, and portfolio management helps to ensure that projects are selected and executed in such a way that they support organizational strategies, goals, and objectives and ultimately realize organizational benefits. The interaction between project, program, and portfolio management enables a number of feedback loops, such as the review and adjustment of portfolios. This in turn informs the objectives of programs and projects, or business impact analysis, to assess the extent to which project deliverables are supporting portfolio value decisions.

Figure 3.1 illustrates the three dimensions of organizational activities, including planning, changing, and operating. While projects, programs, and portfolios can naturally reside across all three dimensions, they tend to be most intense in the changing dimension. After all, the changing dimension can require the most investment of building a viable future, higher risk as there are more uncertainties and will be the potential of higher rewards. The combination of project-, program-, and portfolio management forms the basis of organizational project management (OPM).

OPM is a structure in which project, program, and portfolio management are incorporated with organizational enablers for accomplishing strategic goals and objectives.[2] For organizations, OPM is an important contributor to the achievement of strategic goals and objectives. The purpose of OPM is to ensure that the organization is executing on the right projects, in the right way. Project management is integrated into organizational planning, operations, and change in the form of projects, programs, and portfolios, and the project management office (PMO) is then established to align their execution to the broader organizational strategy. The way in which project, program, and portfolio management each contributes to organizational strategy is unique.

Definition. OPM is a structure in which project, program, and portfolio management are incorporated with organizational enablers for accomplishing strategic goals and objectives.

---

1 Wu, T., 2016. *The Sensible Guide to Passing the PfMP (Portfolio Management Professional Exam)*, iExperi Press, Montclair, NJ. Figure 2. p. 42.

2 Williamson, B. & Wu, T., 2019. *The Sensible Guide to Key Terminologies in Project Management*, iExperi Press, Montclair, NJ. Glossary.

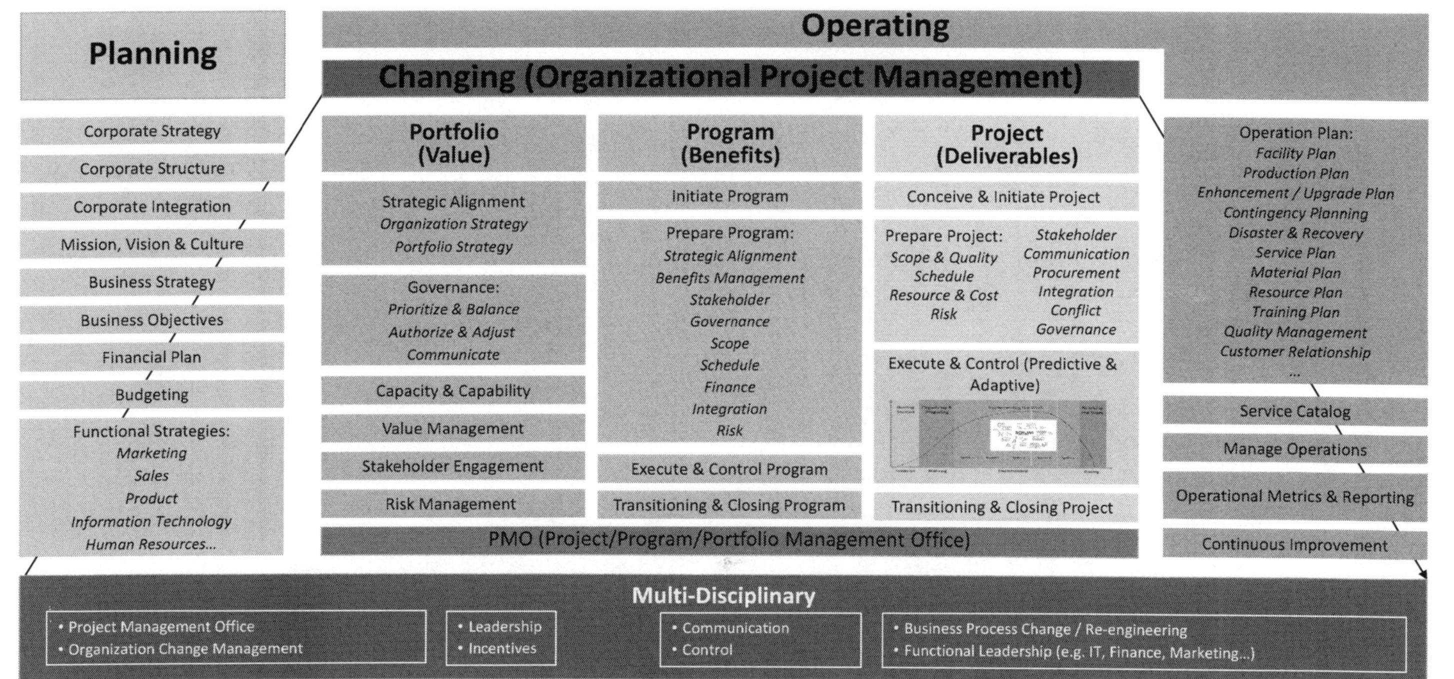

Figure 3.1 Organizational dimensions and selective activities. (Reprinted with permission from PMO Advisory LLC.)

Portfolios and business value. A portfolio is a collection of related company endeavors often encompassing projects, programs, smaller portfolios, and even other activities, such as operations. A project portfolio is similar in concept to a financial portfolio, because it is designed to maximize a desired outcome while also balancing risks. Portfolio management is the application of appropriate management processes, knowledge, tools, and techniques to portfolios, and its primary purpose is to invest organizational assets wisely and to achieve business value. This is important, because in most organizations, resources are limited creating the need for careful investment decisions. Portfolio management is thus the discipline that identifies, selects, prioritizes, approves, governs, and controls the performance of investment projects in line with organizational strategy. Since portfolios should be directly aligned with strategy, portfolio management serves as the bridge that connects planning with execution.

Insights from the front line. Risks are not always negative. Negative risks are called "threats", and positive risks are known as "opportunities". For example, a SAP project often has a risk or threat of not being able to find important solution architects. But if the economy worsens, and many companies are stopping their SAP projects, then it can be an opportunity for the project as more solution architects are available. Also, from a project management perspective, regardless of positive or negative risks, badly managed risks are considered poor project management.

Programs and benefits. A program is a group of highly related and integrated projects, such that the total summation of project results outweighs the simple aggregation of deliverables. For example, a company looking to build a new capability might recognize four related capability-building projects. These projects might be (1) business processes, (2) information systems, (3) regulatory approval, and (4) hiring and training new employees. While each individual project will serve its immediate purpose and objectives, without the overarching rationale from the program (i.e., developing a new capability), the business justification for those projects is nonexistent, or at least weak.

Despite the benefit of programs, the collective interdependence of projects within a program can be much more challenging than individual projects in isolation. To launch this capability, all projects must come together with tight integration of scope, schedule, and process interoperability. Without tight integration, inefficiencies and unforeseen costs can threaten program success. For example, if people are hired and trained prematurely, or the recruitment is initiated prior to receiving the right approval, then there may be no work for the new employees to perform. This would be a waste of the company resources invested into recruitment, selection, and training that yields no return. In order to justify this added complexity, programs should always derive organizational value that cannot be attained when managing projects individually.

Projects and specific results. A project is a temporary and often unique endeavor to produce specific outputs, deliverables, or other results. Project management is the discipline of successfully implementing projects within the organization. In the context of OPM, a project is viewed as the most common unit of work. When the goals and objectives of a project are clearly defined in terms of their strategic contribution, then the successful completion of each project is a milestone achievement toward the broader strategic value of the program and/or portfolio. Thus, while projects can be relatively small unit of work as compared with programs and portfolios, they are important moving parts that enable strategic execution.

PMOs for sustainability of project execution. A PMO is a centralized body designed to provide support and management functions for projects, programs, and portfolios. The term can also refer to portfolio management office and program management office. A *portfolio management office* usually exists at the organizational level and is centered on managing project portfolios. A *program management office* exists at the program level and is assembled to manage large programs. Once a program is complete, these PMOs often disperse. The *project management office* can exist at many levels of the organization (such as department, business unit, or even organizational level) depending on the magnitude and purpose of the project.

However, these terms are not standardized in the industry, and the choice of name often reflects an organization or industry preferences. In the defense industry, for example, "programs" are prevalent, and they are more equivalent to projects in consumer industries. In addition to the naming confusions, PMO capabilities differ widely among the practicing organizations. According to the Gartner Industry Research Report, organizations who establish a PMO with suitable governance and project management standards are likely to experience half the major project cost overruns, delays, and cancelations of those that do not.[3]

For project, program, and portfolio management and PMOs to be truly effective, it is important to optimize and integrate them with other disciplines. Depending on the type of change, project management should work closely with other disciplines such as organization change management, process redesign, and leadership development to maximize the effectiveness of the endeavor.

Insights from the front line. In PMO Advisory's consulting methodology, there are four levels of PMOs as shown in Figure 3.2.

1. The first level is the Learning PMOs, and this is often an organic outgrowth as company tackles more projects and there are more project professionals. These professionals would likely wish to share experiences and battle stories, learn from each other, and collect and utilize

[3] PM Solutions Research., 2010. *The State of the PMO 2010*. Research report. PM Solutions, Glen Mills, PA.

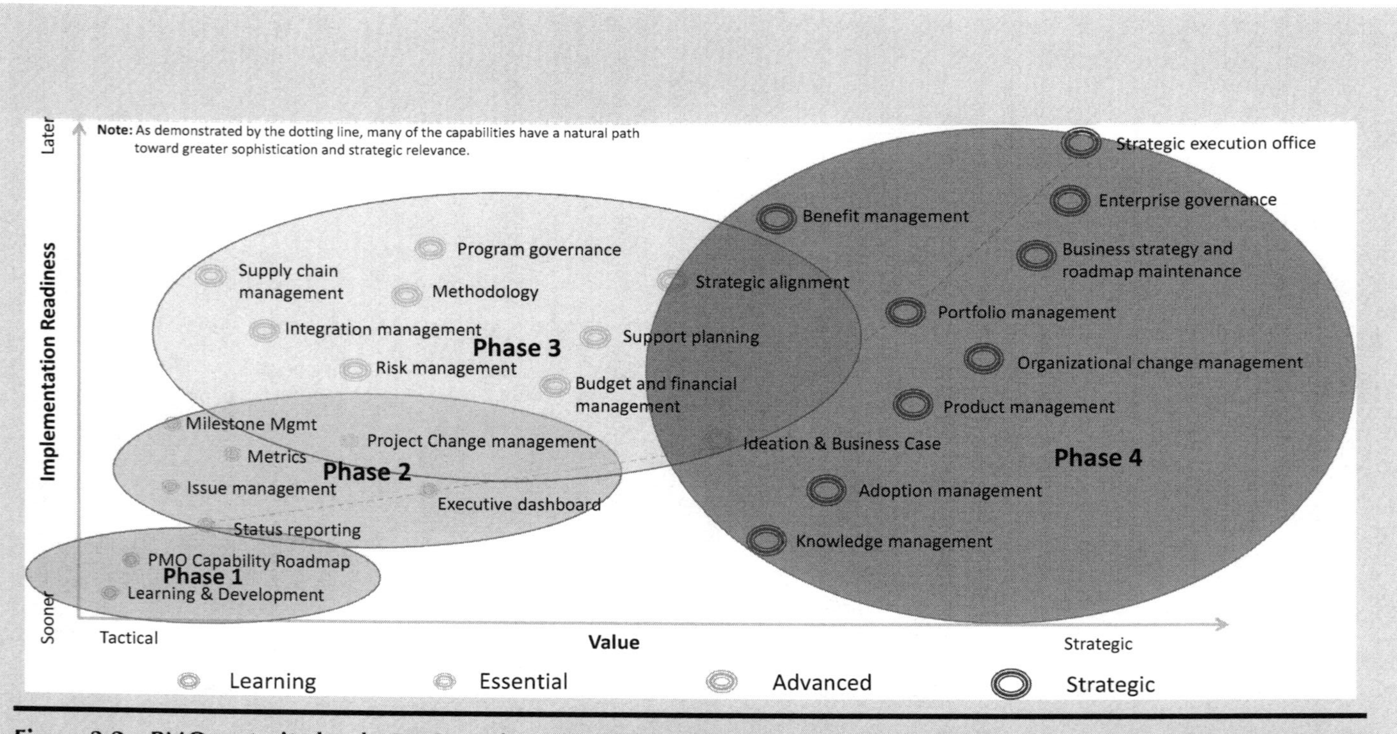

Figure 3.2 PMO maturity levels. PMO, project management office. (Reprinted with permission from PMO Advisory LLC.)

good practices. By creating a community to share, learn, and train, the Learning PMOs are formed. Sometimes, these are not formal structures and can be viewed as a "club" or "community" group within the organization.

2. The second level includes *Essential* PMOs, also known as Controlling PMOs. These are created primarily to monitor, report, and control portfolios of projects and programs where strict reviews and adherence to governance requirements are a priority.
3. *Advanced* PMOs, sometimes referred to as Supportive PMOs, focus on enhancing execution capability and providing decision support. This is achieved through best practices, expertise, the provision of information, and other tools.
4. The final level, *Strategic* PMOs, focuses on business or organizational value and planning for the long term. These PMOs can also be referred to as Directive PMOs, because they are often focused on directing the allocation and utilization of resources. Due to these differences, there is currently no standard body of knowledge on PMO. For simplicity's sake, in this book and unless otherwise stated, "PMO" refers to a project management office.

3.2 Entrance and Exit Criteria for Ideation

As the first phase in the project life cycle, ideation may or may not have a formal starting point. In organizations with formal processes for gathering ideas and strategy making, ideation begins when there are credible suggestions from employees and customers, insights from various analysis such as the Strength, Weakness, Opportunity, and Threat (SWOT) analysis or enhancements to existing products or services, current programs and portfolios in which projects are specific components, and residual or new ideas from an earlier time. In less structured organizations, ideas can come about at any time and often in an ad hoc manner.

Regardless of whether organizations have formal or information ideation processes, these ideas are then reviewed and vetted by executives, whether in teams or by individuals. In a structured system, there should be processes to

- analyze the ideas and evaluate its importance,
- group ideas into various categories and sometimes by project portfolios,
- develop business cases to support the advancement of ideas, and
- prioritize by a range of factors that are importance to the sponsoring organization.

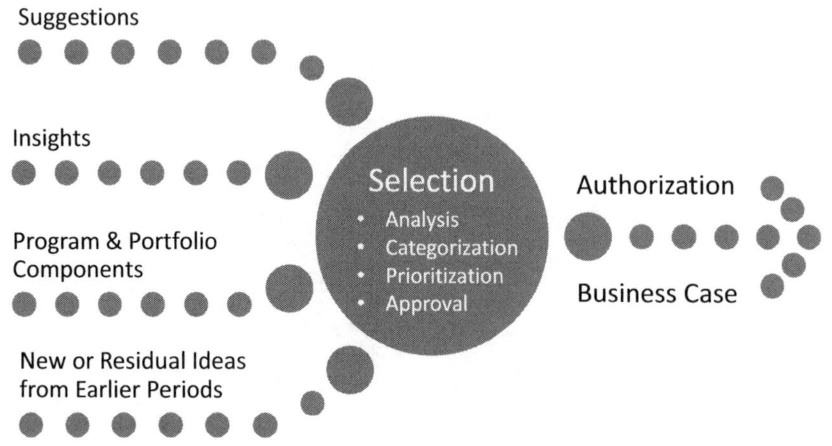

Figure 3.3 Entrance and exit criteria for Ideation. (Training and consulting content from PMO Advisory LLC. Reprinted with permission.)

Figure 3.3 highlights the key ideation processes. The phase concludes when the ideas and the associated business cases are approved for implementation, declined, and thus closed, or deferred to another time.

3.3 Project Selection

Peter Drucker once said that "Nothing is less productive than to make more efficient what should not be done at all". Similarly, doing a project "right" is not enough if one is not doing the "right" project. Organizations have limited resources available to transform ideas into action. This means that projects must be carefully selected such that their implementation (and the resources utilized during implementation) is measurably beneficial to the organization. Projects are selected based on the value that they bring to the organization, the organizational problems that they solve, the organization's ability to implement projects successfully, the accessibility of resources required to implement them, and support from internal sponsors or external stakeholders. Ultimately, the projects selected for implementation should be those that best enable the organization to achieve its strategic purpose.

Figure 3.4 outlines the process of selection. Ideas, or strategic problems, are assessed according to their feasibility (likelihood of achieving success despite risk and variability) and benefits (magnitude or impact of the project deliverables). Those ideas that are ranked the highest on feasibility and benefits are chosen to move forward in the selection process. Business cases are then used as a practical

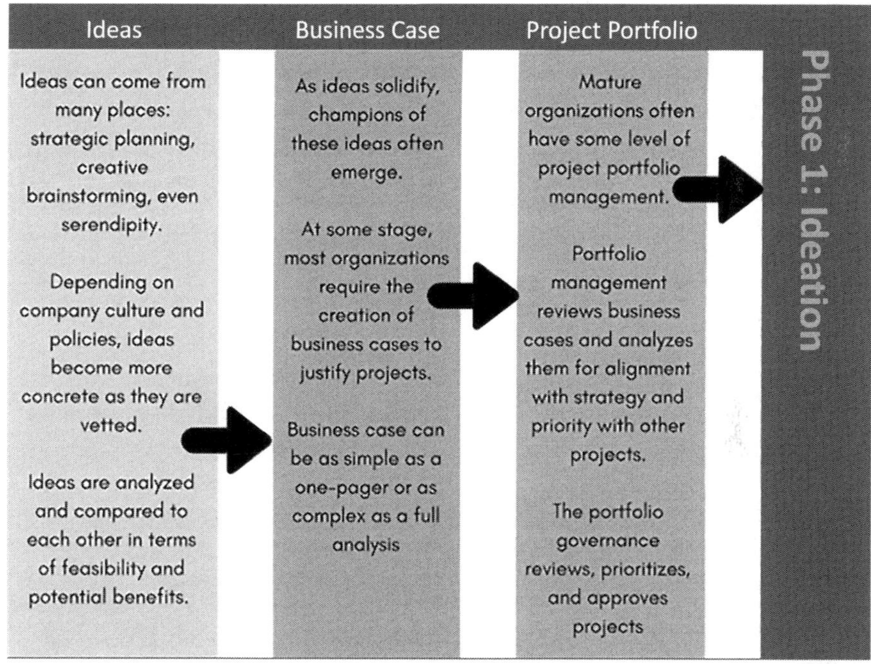

Figure 3.4 Project life cycle phase 1 – Ideation. (Training and consulting content from PMO Advisory LLC. Reprinted with permission.)

analysis of a business problem and provide helpful information on the feasibility of options for addressing the problem and a recommended course of action to derive the maximum benefits.[4] In mature organizations with portfolio oversight, the portfolio manager(s) will review business cases based on potential value to the organization, customer interests, strategic alignment, and/or human and financial resources.[5] Those projects that are successfully selected from their business case will form part of the broader portfolio of projects and programs. Criteria used for selecting ideas and business cases can include financial criteria such as return on investment, net present value, internal rate of return, benefit–cost ratio, and payback period, and nonfinancial criteria such as corporate social responsibility, customer

[4] PMBOK Guide (6th ed.) Project Management Institute, 2017. *A Guide to the Project Management Body of Knowledge (PMBOK® Guide) (6th Ed.)*. Project Management Institute, Newtown Square, PA.

[5] Project Management Institute, 2013. *The Organizational Project Management Maturity Model (OPM3®) - 3rd ed.* Newtown Square, PA.

satisfaction, improved brand value, improved sustainable competitive advantage, and market share increase.

Even when organizations have all the necessary information to select the most feasible and beneficial projects, and even when they follow a logical process of selection, making sound project selection decisions is not without its challenges. Often selection is clouded with too little, too much, or outdated and incorrect information. Furthermore, *people* make project selection decisions, and this means that the process is often influenced (positively or negatively) by emotions, passion, experience, and intuition. Some organizations suffer from an overemphasis on speed of selection decisions, and others are so slow to select that "analysis paralysis" ensues and projects are too late to initiation. Given the constantly changing organizational environment, selection decisions are often made with a level of uncertainty and bias, as decision-makers balance multiple, complex, contributing factors.

3.4 Making Sound Project Decisions

Organizations operate in a complex and changing environment where decisions like project selection need to be made in a timely, yet systematic, fashion. Organizations may make use of different selection criteria that are tailored to organizational needs and stakeholder benefits. However, whatever the criteria, a systematic approach to project selection decisions is critical for minimizing uncertainty and bias. There are a number of different approaches to managing the challenges of project selection and making sound selection decisions. One of the decision-making techniques receiving attention today is the Analytical Hierarchy Process (AHP).[6] The AHP is a mathematical process that focuses on organizing and analyzing complex, multicriteria decisions by decomposing decision problems or goals into a hierarchy of smaller components. By focusing on the smaller components and making systematic pair-wise comparisons, decision-makers are able to make clear distinctions and convert subjective evaluations into objective, quantifiable scores. These scores are then used in quantitative analysis and aggregated back into the larger problem. There are a number of ways to approach AHP, including the multicriteria scoring model (Figure 3.5), multicriteria weighted ranking (Figure 3.6), and single-criterion prioritization model (Figure 3.7). The AHP is a robust and adaptive technique most suitable for complex, criteria-based business scenarios. The advantage of this technique is that it is intuitive and simple to establish, communicate, train, and use.

A detailed discussion of AHP is beyond the scope of this book, and readers are encouraged to read Thomas Satty's 2012 book entitled "*Decision Making for Leaders:*

[6] Vargas, R. V., 2010. *Using the Analytic Hierarchy Process (AHP) to Select and Prioritize Projects in a Portfolio.* Paper presented at PMI® Global Congress 2010—North America, Washington, DC. Project Management Institute, Newtown Square, PA.

BEST PRACTICE EXAMPLE 1:
Multi-criteria Scoring Model

Based on the weighted score across 7 criteria, Component A is the highest priority component, followed by Component C and B.

		Component A		Component B		Component C	
List of Criteria	**Weight**	**Raw Score**	**Weighted Score**	**Raw Score**	**Weighted Score**	**Raw Score**	**Weighted Score**
Financial Metrics	20%	High	2.00	Med	1.00	High	2.00
Marketshare Gains	20%	High	2.00	Med	1.00	Med	1.00
Competitive Advantage	15%	Med	0.75	High	1.50	Med	0.75
Customer Satisfaction	15%	High	1.50	Med	0.75	High	1.50
Implementation Feasibi	15%	Med	0.75	Low	0.00	Low	0.00
Ease of Market Entry	10%	Low	0.00	High	1.00	Med	0.50
Timing (Why now?)	5%	Med	0.25	Med	0.25	High	0.50
Total Score:			**7.25**		**5.50**		**6.25**

Raw Score: Low = 0; Medium = 5; High = 10

Figure 3.5 AHP multicriteria scoring model. AHP, Analytical Hierarchy Process.

The Analytic Hierarchy Process for Decisions in a Complex World, New Edition". The following are some "common good practices" to consider when implementing the AHP in project selection decisions:

1. The decision-making process should be agreed upon in advance, and the full process should be communicated to all stakeholders. This is perhaps the single most important step in the process. By agreeing with the process in advance, participants are aware of the ground rules, procedures, and major factors of consideration.
2. All stakeholders should be trained to focus on tasks, issues, and objective data, rather than personalities or positions. Since AHP is often used on complex issues, which can be emotionally charged, it is important for the participants to focus on the problems and not personalities. Whenever possible, supporting their ideas with quantifiable and objective data is important.
3. The overall objectives of the selection process (i.e., the strategic selection of projects) should be prioritized over narrow, immediate, and subjective interests. For portfolio managers managing the process, it's imperative to focus on the ultimate business or organizational objectives of the portfolio, not a particular project or program.

BEST PRACTICE EXAMPLE 2:

Multi-criteria Weighted Ranking

Projects in the Portfolio are ranked based on criteria, highest priority is assigned to the project with the lowest score, calculated as an average of the ranking.

Projects	Criterion 1		Criterion 2		Criterion 3		Criterion 4		Priority	
	Measure	Rank	Measure	Rank	Measure	Rank	Measure	Rank	Score	Priority
Project 1	32.0	2	17.6	2	8.8	2	$4M	1	1.75	1
Project 3	28.0	4	37.8	1	18.9	1	$5M	2	2.00	2
Project 4	31.0	3	16.9	3	8.45	3	$6M	3	3.00	3
Project 2	38.0	1	11.9	4	5.95	4	$8.6M	4	3.25	4
Project 5	20.0	6	10.8	5	5.4	5	$9.2M	5	5.25	5
Project 6	24.0	5	4.2	6	2.1	6	$10.40	6	5.75	6

Figure 3.6 AHP multicriteria weighted ranking. AHP, Analytical Hierarchy Process.

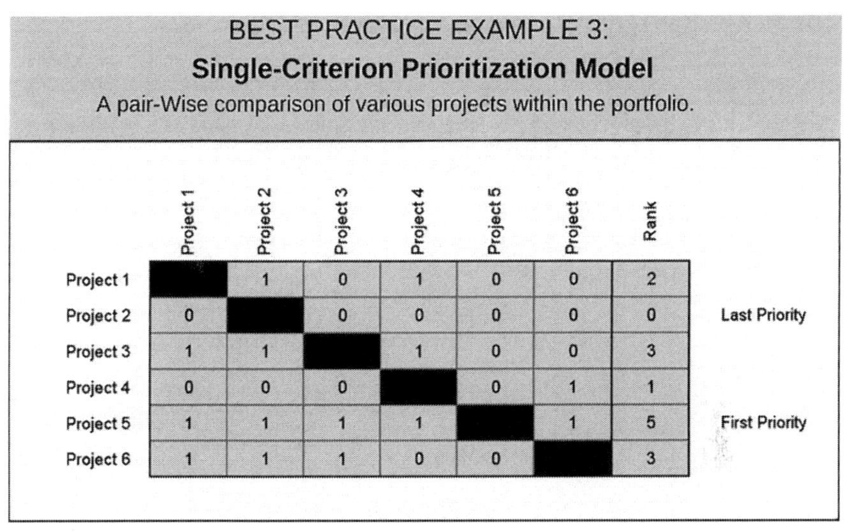

BEST PRACTICE EXAMPLE 3:
Single-Criterion Prioritization Model
A pair-Wise comparison of various projects within the portfolio.

	Project 1	Project 2	Project 3	Project 4	Project 5	Project 6	Rank	
Project 1		1	0	1	0	0	2	
Project 2	0		0	0	0	0	0	Last Priority
Project 3	1	1		1	0	0	3	
Project 4	0	0	0		0	1	1	
Project 5	1	1	1	1		1	5	First Priority
Project 6	1	1	1	0	0		3	

Figure 3.7 AHP singlecriterion prioritization model. AHP, Analytical Hierarchy Process.

4. The application of the decision-making process should also be consistent and uniform, without being dogmatic. This means removing politics and embracing common ground, while also discussing differences openly and respecting the outcomes.
5. Finally, depending on the specific AHP method, it is important to agree on the relative weight of the factors and scoring before using actual project ideas. By focusing on the objective criteria and the relative emphasis largely based on strategic objectives, participants are more likely to view the bigger picture, the moment the focus shifts toward specific initiatives, politics, and interests, which can come into play.

3.5 Developing a Culture of Execution

No business culture is the same. Business culture is a unique combination and interaction of shared values, beliefs, behaviors, and implicit workplace rules. We cannot touch or hold business culture, and yet we know that it is there. Culture is the invisible energy and the unspoken way of communicating, behaving, and feeling between groups of people that gives organizations their unique characteristics. Project teams and processes are often sensitive to broader business culture or take on a culture of their own, particularly when projects are global. As technology breaches geographical borders, so the need for project managers to manage international projects is increasing. But multicultural project teams often accompany international projects, where each team, from each geographical workplace, holds a different set of values, beliefs, and behaviors. This makes it difficult to coordinate a single project culture of execution. Some of the most common cross-cultural dimensions of global projects include the following[7]:

- **Power distance.** The inequality of power across project and organizational structures and hierarchies, which become particularly complex when project stakeholders span across companies and/or geographical areas. Flat structures typically have low power distances, and tall structures have high power distances.
- **Uncertainty avoidance.** The level of comfort with, or acceptance of, uncertainty and risk. Different countries, and even organizations, usually have characteristically low or high levels of uncertainty avoidance. The higher the uncertainty avoidance, the more time and resources an organization or project will typically require for adjusting and changing.
- **Achievement orientation.** The desire for action, results, and excellence. Those personalities with high achievement orientation tend to take initiative and prefer to be consulted, rather than follow instructions. Project managers are often high on achievement orientation.
- **Individualism.** The degree to which people prioritize their own achievements against the achievements of the team. High levels of individualism emphasize a focus on individual team member inputs and results, and low levels of individualism emphasize a focus on project team input and results.
- **Context.** The information around an event, and the nature of its communication. High-context cultures have access to multiple sources of information and view communication as an engagement using expression and social cues. In low-context cultures, more information is usually required before a decision can be made.

Definition. Culture is a system of values, behaviors, attitudes, and traditions that are often unspoken in an organization, which affects the planning and execution of projects, programs, and portfolios.[8]

---

[7] Anbari, F. T., Khilkhanova, E. V., Romanova, M. V., Ruggia, M., Tsay, H.-H., & Umpleby, S. A., 2009. *Managing Cross Cultural Differences in Projects.* Paper presented at PMI® Global Congress 2009—North America, Orlando, FL. Project Management Institute, Newtown Square, PA.

[8] Williamson, B., & Wu, T., 2019. *The Sensible Guide to Key Terminologies in Project Management*, iExperi Press, Montclair, NJ. Glossary.

The project manager is instrumental in controlling these dimensions and leveraging business culture as a tool for successful execution. Cultures that embrace project management and execution recognize and understand the value of project management, engage executive level project sponsors, and ensure a strong alignment between strategy and projects.[9] Projects that are executed within a project-oriented culture are more likely to have improved performance in project management. This means that, depending on how it is managed, culture can act as an obstacle or a catalyst for high-performance projects. A culture of project excellence is focused not just on deciding *what* to do but also on following through with action. High-performing project teams have strong and established project execution cultures where members implicitly know what to do and how to do it. In this culture, existing project members are resilient to negative influence from new members and act as examples of the culture. Here, executive and management level leadership sets the tone for successful execution.

In order to leverage culture for project effectiveness, project managers must create a project vision for the end state so that all team members understand what they are working toward achieving. A clear approach for the project and for effecting change should then be developed to guide the project team to the end state. During the project life cycle, constructive attitudes, behaviors, and results should be rewarded with a focus on collaboration, rather than individualism.

3.6 Business Case

One approach to develop a culture of execution excellence is the enforcement of creating rigorous business cases, and level of analysis and supporting information should be proportional to the size of investment, complexity, and the consequence of failure. A business case is "a feasibility report for projects, programs, and portfolios used to establish the rationale for undertaking the endeavor".[10] As a feasibility report, it's imperative for a business to clearly specify the benefits and value gained from completing a project. But this benefit must weight against the costs. A common problem with business cases is the overstatement of the benefits and an underestimation of the cost, resources, schedule, and other project considerations. Using these gimmicks will artificially boost the case but may ultimately lead to failed expectations. Project professionals should consider these project management knowledge domain questions listed in Table 3.1 during the Ideation Phase, especially as a part of developing business cases.

---

[9] PMI Pulse of the Profession: Capturing the Value of Project Management, 2015.

[10] Williamson, B., & Wu, T., 2019. *The Sensible Guide to Key Terminologies in Project Management*, iExperi Press, Montclair, NJ. Glossary.

Table 3.1 Project Ideation Phase

#	*Knowledge Domain*	*Key Question During Ideation Phase*
1	Integration management	Are there other projects (or programs and portfolios) dependencies on a proposed project? What is the best approach to execute this project?
2	Stakeholders management	Who are the key stakeholders, especially the supporters and dissenters?
3	Scope management	What will project deliver and how does it fit with the overall business environment and strategy?
4	Schedule management	Are there strict deadlines for this project? If yes, do we have reasonable amount of time to undertake this project?
5	Resources management	What are the most crucial resources required for this project? Do we have access to these resources?
6	Cost management	How much funding are we willing to invest? Does it make financial sense?
7	Communication management	Are there special communication challenges?
8	Risk management	How risky is this project? What is our organization's risk appetite?
9	Quality management	Are there special quality considerations? What are the minimal acceptance thresholds?
10	Supply chain management	Are there are external resources required for this project?
11	Conflict management	Are there known conflicts on this project? What is the intensity of those conflicts?
12	Governance management	Are there special governance considerations?
13	Issue management	What are the issues now that's already impeding progress?
14	Adoption management	Are there people and organization change implications? What's the initial perception on the degree of change?
15	Operation management	What happens after the project completion? If there are operational considerations, to what extent will that impact the current environment?

3.7 Project Management in Motion

This book is intended for a wide range of audiences and across various industries and functions. Therefore, to balance between competing demands of ease of use, comprehensiveness, modularity, flexibility, and upgradeability, the author has adopted a modular format to this book. This includes the relatively short chapters on key topics. In addition, the appendices include three vital sections.

- Appendix A contains a list of commonly used project management templates for both predictive and adaptive project management approaches.
- Appendix B contains an integrated case study based on a fictitious global company undertaking a number of projects and confronting various project challenges. The case maps closely to the chapters and sections, key concepts, and relevant tools of this book. The goal is to provide instructors, students, and practitioners with a realistic project example to practice and apply the key learnings in the book. For more information and additional case studies in which the author plans to create over time, including potential collaboration opportunities, visit www.optimizepm.com.
- Appendix C contains a glossary of selective terms, reprinting with permission from Dr. Te Wu and Mr. Brian Williamson's book titled "*The Sensible Guide to Key Terminologies in Project Management*", iExperi Press, Montclair, NJ.

Chapter 4

Initiation – Starting the Project Right

Summary

Continuing the project management journey, this chapter discusses the next phase of project life cycle – Initiation. The primary purpose of the Initiation Phase is to determine the optimal approach to execute the project and to ensure the continual support of the organization, after conducting a more thorough evaluation on the implementation feasibility of the project. The attention of the organization shifts from that of business and organizational planning to execution. Often, project professionals and selective core team members, who are often the subject matter experts, are first engaged here. With their combined intelligence evaluating the approved ideas, organizations gain greater confidence in the viability of the project. On the flip side, if the project implementation is more difficult or cost or time-consuming that was approved in the business case, organizations can use this opportunity to reconsider moving forward with the project.

During this phase, project professionals utilize a wide range of assessment and evaluation tools, such as stakeholder identification and assessment, risk identification and evaluation, initial requirement analysis, and estimation of scope, resources, and cost. For selective projects, there can be further analysis on organizational readiness and adoption, evaluation of conflicts including politics, and impact on operations. The cumulation of this work is a project charter, a contract between the sponsor, representing the customers and executives, and the project manager, representing the project team.

This chapter addresses these fundamental questions:

1. What are the key questions to tackle during project initiation?
2. What is the problem of fuzzy front end? How to get started?
3. How to manage the fuzzy front end?
4. What is a project charter and why is it important?

4.1 The Challenges of Getting Started

Upon entering the Initiation Phase, project managers confront a wide range of challenges, from the basic questions of how to get started to more advanced questions such as estimating cost, approximating schedule, and identifying resource needs. Project professionals are encouraged to ask these important questions and discover answers and solutions especially on challenge projects as answers to these questions are essential for success. Table 4.1 highlights some questions to consider during the Initiation Phase.

This book will address these and other questions throughout the book, but mainly in the Part 3: Knowledge Domains.

Table 4.1 Project Initiation Questions

#	*Knowledge Domain*	*Key Question during Initiation Phase*
1	Integration management	What are the main project dependencies? What approach should we utilize in the project execution? How to manage change? What are the broader project considerations, especially if the project is a part of a program or portfolio?
2	Stakeholders management	Who are the key stakeholders? What are their interests, either at the beginning, during implementation, or after the project is completed? How much power to they hold? What are their criteria for success?
3	Scope management	What are the key requirements of the project? Who will be using or benefiting from this project? How will the project deliverables be used?
4	Schedule management	What is the allowable duration of the project? Is there a strict deadline? If yes, why? Are there any major dependencies, especially if the project is a part of a program or portfolio? What are the important milestones?

(Continued)

Table 4.1 (*Continued*) Project Initiation Questions

#	*Knowledge Domain*	*Key Question during Initiation Phase*
5	Resources management	What are the broad resources needs for this project, people, and nonpeople resources? Are these readily available internally or do we need to acquire them from external sources? For people, what are the technical and behavioral skill requirements for this project? Are there scarce resources, so scarce that without them the entire project may be in jeopardy? Are the team colocated or virtual?
6	Cost management	What are the sources of funding? What are the cost constraints? Is there a predetermined budget, for example, from the business case? How feasible is the budget? Is there flexibility to ask for more money or return unused funds? How to secure the required funding?
7	Communication management	Are there communication challenges? If yes, what are they? What are the communication tools available (e.g., e-mail, intranet/Internet, wiki, social media, phone, web conferencing, etc.)?
8	Risk management	How risky is this project? What are the project risks? Who can help the project team identify and evaluate the risks?
9	Quality management	What are the quality considerations? How do the quality requirements impact scope?
10	Supply chain management	Should we "make" or "buy" selective resources? Does the project require external resources? If yes, are suppliers for our needs commonly available or rare and difficult to find?
11	Conflict management	Are there conflicts of personality, process, and priority on the project – currently or in the future?
12	Governance management	Who are the decision-makers and key influencers? Is it an individual or a team of people? What are the key governance processes? How many layers of governance for this project?
13	Issue management	What are the issues that are already confronting the projects? How to tackle these issues?

(*Continued*)

Table 4.1 (*Continued*) Project Initiation Questions

#	*Knowledge Domain*	*Key Question during Initiation Phase*
14	Adoption management	Are there organizational and people implications, intended or unintended, from this project? What is the degree of impact? Does the project need a specialist dedicated to manage adoption?
15	Operation management	Will there be a need to manage the project deliverables after the completion? If yes, what are the degree of impact to the operational environment?

For now, let's tackle a common early challenge – how to get started as efficiently and effectively as possible. Naturally, the bigger, the more complex, the greater the number stakeholders, the higher the risk and uncertainty; the more innovative the project deliverables, the challenges of determining that optimal start becomes greater. Often there are many, if not infinite ways of getting started. What should be the first step? This is the problem commonly known as the "fuzzy front end". No wonder the Initiation Phase is often considered the most critical phase of these complicated projects. Without having a clear awareness of these challenges, projects can start off poorly, which can endanger the overall success of projects.

4.2 What Is the Fuzzy Front End?

The Chinese have an old proverb, "a 1000-km journey starts with the first step". But what is the first step? How do you know it's in the right direction? What are the prerequisites that must be dealt with? What if the destination itself is rather fuzzy?

Projects, especially complex projects with high uncertainty as in the case of strategic innovation, often have many if not infinite approaches of addressing the projects.[1] One of the first and perhaps most challenging project management activity is to find the optimal approach addressing this challenge. Seasoned project managers must analyze the situation that includes the sponsoring organization(s), stakeholders, capability and maturity of the processes, the project deliverables themselves, and a myriad of other environmental, organizational, and team factors to determine the optimal approach.

[1] Mootee, I., 2011. *Strategic Innovation and the Fuzzy Front End.* Retrieved from https://iveybusinessjournal.com/publication/strategic-innovation-and-the-fuzzy-front-end/ on February 25, 2019.

4.2.1 Anatomy of the Fuzzy Front End

The fuzzy front end is defined as the period between the project initiation as an idea and when it is ready for planning. In this critical Initiation Phase, key customer problems, issues, process gaps, untold opportunities, future upgrade plans, and friendly chitchats should be identified. They may be not formally documented previously, but now it is time to evaluate them systemically. Furthermore, building strong customer relationship starts here with the potential of building goodwill and marching toward project success. At this very beginning, customers are usually in a messy state of mind, on what they want (requirements) and how to approach the implementation as there can be many ways. At this early phase, project professionals can help their customers and sponsors by analyzing the various options and carve a path forward. There are potentially endless lists of factors contributing to and aggravating the "fuzziness" of the projects at the Initiation and Preparation Phases. Below are three main categories of factors:

1. Project factors
   - Project deliverable itself is poorly defined or understood (e.g., technically difficult to specify the details or lacking internal expertise to define the deliverables fully) or pin down (e.g., as there are multiple conflicting factors)
   - Innovative projects, especially those at the bleeding edge, are by definition implementing new and often unproven technology. This creates additional uncertainties such as stability, quality, functionality, and potential integration issues with the rest of the systems and processes.
   - Project deliverable can be changing, whether the reason for the change is environmental such as competitive moves or changing regulation, and/or internal such as stakeholders switching their preferences or disagreements
   - Project constraints can also contribute to this challenge. As constraints tighten, the implementation challenge becomes immense, and at some point, beyond the breaking point.
2. Organizational factors
   - The project team's maturity and cohesiveness are important factors to consider. With a high-performing team with a strong understanding of project management, most of the lesser challenges would essentially go away through superior problem-solving and teaming skills. But the same issues confronting a less mature team would often result in bigger obstacles.
   - The sponsoring organization's culture, processes, and competencies are important considerations. The "fitness" and "alignment" of decision-making and governance can have a major impact. For example, if key decisions take weeks or months to make, then iterative approaches like Scrum are likely to be suboptimal and directly impact the time required to make key decisions at the start of the project.

  - The maturity of the project management and the appropriateness of methodology can also be an important factor. If a team that excels in traditional project management is coerced to be more Agile, especially without the support of the underlying processes, then what likely happens is a hodgepodge of approaches creating chaos and confusion.
  - Finally, all organizations have some level of politics. Unless they are well managed, conflicts of interest and priorities often appear at the onset of projects, delaying and possibly even killing projects.
3. Environmental factors
  - Environmental factors are considerations external to the project and the organization working on the project, which on a large and complex project can be quite numerous. For well-defined projects in which the underlying technology is mature, the environmental impact can be well estimated and managed. But for projects that are truly innovative or with a high degree of competition, environmental impact can be more significant.
  - Examples of environmental factors include the following:
    - Social economic – for example, society's acceptance of artificial intelligence in driverless vehicles, even if the technology is proven. Or, the state of the economy.
    - Industry and competitive – the amount of competition in an industry.
    - Regulatory – governmental involvement, regulations, guidelines, and even industry standards can impact how projects are defined and thus create additional concerns at project initiation.

Activities in the fuzzy front end are usually difficult to anticipate, understand, and determine. Some project managers underestimate the seriousness of fuzzy front ends. Their attitude may be to underemphasize this planning since it's so chaotic. They may not have a vision for the project flow or create a full project plan to understand deeply customer needs and requirements. Some may even hide behind the "Agile" methodology so there are excuses for weak analysis and little documentation; everything "goes" attitude toward scope definition and ultimately a poor foundation for the project. This can surely cause serious gaps and misunderstandings of customer needs. Accordingly, potential risks and issues are not identified well. And project delays, risk mitigation alternatives, and resource allocation are not done properly either. As a result, chances of project failure and customer dissatisfaction are high.

4.3 Good Practices for Managing the Fuzzy Front End

With often many choices to make at the onset of projects, project managers should drive to a degree of certainty as quickly and as efficiently as possible. Table 4.2 contains ten good practices to consider to drive toward stability and certainty.

Table 4.2 Good Practices for Getting Started

Good Practice	*Description*
Vision	Develop a clear direction and compelling future for the outcome of the project. Having a strong vision anchors the project team toward a common set of objectives.
Champion	Seek one or more change champion(s) who can engage key stakeholders, develop them into willing and collaborative partners, drive them toward achieving the vision, and fight the good battles.
Sponsorship	Make sure there is executive management involvement that provides coverage on contentious issues. For larger projects, consider establishing robust governance to guide key decisions.
Customer	Involve customers and end users early on, to refine product ideas and project implementation processes.
Process	Identify and implement a project management approach early on, to instill process rigor early. But also be willing to change as innovative projects can have many detours.
Collaboration	Perform outreach activities with key internal resources, such as functional organizations, to create a dynamic environment for idea exchanges and mutual collaboration.
Lessons learned	Examine important lessons from the past similar endeavors and plan to address them in current projects. Whenever possible, seek expert advice from the previous project manager and other important stakeholders.
Prioritization	Innovative and complex projects often have a long list of desired project outcomes. Spend time early on to organize, categorize, and prioritize the wish list items. Draw a firm line between the essential components versus more discretionary features.
Alignment	Examine the consistency and agreement of the project goals and outcomes with the product objectives and organizational goals. Aligning goals are vital, especially when making intricate trade-off decisions.
Communication	Project managers easily spend 80% or more of their time on communication. A good practice is to conduct "kick-off meetings" as frequently as required to orient the team, onboard new members, and communicate the goals, challenges, and the work remaining to complete the project satisfactorily.

4.4 Seven Steps to Managing the Fuzzy Front End

With often many choices to make at the onset of projects, project managers should drive to a degree of certainty as quickly and as efficiently as possible. See Table 4.3 for seven good practices to consider to drive toward stability and certainty.

Table 4.3 Good Practices to Manage Fuzzy Front End

#	*Name*	*Description*
1	Evaluate project situation	Evaluate both internal and external considerations surrounding the project, organization, and environmental factors. Internal considerations are largely focused on the immediate and direct considerations such as customers, team capabilities, resources, budgets, and so on. External considerations are the environmental factors that are beyond the PM's control but still have an influence. Develop an early perspective on the challenges ahead.
2	Determine project success criteria and consider developing three sets of success criteria	a. Minimum threshold for success – This is the bare minimum threshold for success. In general, project manager should develop plans that have 85% or higher confidence of achieving this minimum threshold, even though the specific percentage depends on organizations and their risk tolerance. For example, the minimum threshold for successful adoption of the new technology is achieving a net promoter score (NPS) of 25 (on a scale of –100 to +100) with an average of 3 or higher (on a scale of 1–5) for the subcomponent scores. b. Realistic criteria for success – This is the most realistic expectation for success. Project managers should build their project plans to achieve this. In general, project managers should have 75% or higher confidence of achieving the realistic criteria. The realistic NPS is 50 with none of the subcomponent scores below 3 with an average of 4 (on a scale of 1–5). c. Stretch goals or wildly successful – This is by definition "stretched", and generally, it should be subdivided into smaller components. For example, the satisfaction on training is 4.5, exceeding the realistic expectation of 4.

(Continued)

Table 4.3 (*Continued*) Good Practices to Manage Fuzzy Front End

#	*Name*	*Description*
3	Collect customer/ sponsor insights	Work with customers, sponsors, and other key stakeholders to understand their goals and expectations, challenges and hesitancies, and opportunities for improvement. More important, the very act of engaging customers in a professional and sincere fashion builds rapport and trust. Use this opportunity to evaluate and refine the three key factors outlined above: project, organization, and environment.
4	Develop strategic foresights and prioritize requirements	Projects that have severe fuzzy front end problems are nearly always complex, difficult, innovative, and probably sizeable and political. This means that there are considerable conflicts and disagreements, especially at the start of projects. Project managers should work with the key stakeholders and address these conflicts openly, transparently, and collaboratively to achieve agreements or consensus. Ideally, the project manager and even the sponsor are neutral with the primary interest of successfully completing the project.
5	Determine implementation approach (e.g., waterfall, Agile, process rigor)	Based on the earlier analysis, identify the most suitable implementation approach that will help the project succeed. A good implementation approach must consider the three key factors discussed earlier: project, organization, and environment. Depending on organizational and project management maturity, it may be possible to develop a tailored method for the project. Once the method is determined, develop a comprehensive project plan based on engaging and collaborating with the core team and stakeholders.
6	Project visioning	For projects that are ambiguous, controversial, adaptive, or complex, especially when there are major organizational change implications, project managers should consider establishing a common vision for the project team and using this opportunity to bring out the underlying issues and concerns.

(*Continued*)

Table 4.3 (*Continued*) Good Practices to Manage Fuzzy Front End

#	*Name*	*Description*
		Achieving a common project vision may require an investment of time at the beginning of the project, but once the key stakeholders are onboard by contributing to a shared vision, the actual implementation may be significantly faster.
7	Create project charters	Before taking the plunge and diving deep into project planning and organization, project managers should summarize the key findings and understandings, including revision to the business case, and present it to the project sponsor for approval before committing more resources. One of the best practices is to create a project charter, which will be explained in the next section.

Tool: Project Visioning Tool. See Template 3: Project Visioning Tool on page 268 in Appendix A for a detailed procedure and simple template on creating a project vision. Project visioning often occurs on adaptive projects, equivalent to project charter. However, on larger projects with greater stakeholder involvement, project visioning can be performed for any time of projects and in conjunction with project charters.

4.5 The Project Charter

A project charter is "a document issued by an individual or group responsible for sponsoring or initiating the project. The project charter grants formal authorization to the project manager to guide and oversee activities within the context of organizational, contractual, and 3rd party resources." This is often the first major and most important project management document outlining the purpose of the project, describing the project's importance and alignment with organizational strategies, and providing broad boundaries of the project such as cost, schedule, and scope.

Ideally, project managers are assigned early in the Initiation Phase; they can lead or participate in the creation of the project charter. Upon agreement, the project charter can serve as a documented agreement between two or more parties (e.g., sponsor or customer and the project manager representing the project team), usually outlining some sort of value exchange. For example, the project sponsor or customer might agree to obtain the project deliverables upon completion of the project and in return to compensate the project managers and their teams for the project.

The project charter is often a working document that adapts as project conditions change. An updated version of the project charter should always be widely shared among all project stakeholders in order to maintain a common understanding of mutual expectations, resource allocations, the role of the project manager, and the link between the project and strategic objectives. In this sense, the project charter serves as a communication tool to align the project team. The project charter is also often used as the exit criteria for the Initiation Phase of projects.

4.5.1 Project Charter Key Elements

The project charter document can include high-level and detailed components, such as project synopsis, initiative type, associated programs (if any), roles, project manager, program sponsor (if any), program manager (if any), business objectives, business case summary, etc. The following provides a short description of each component:

- **Project name.** Name of the project as agreed with the project sponsor and project core team members.
- **Project description.** A brief overview of the project. This includes compliance and regulatory requirements of the project, for example, a specification that the project be launched by a certain date. If the project is part of a larger program, then it is also indicated under the project synopsis section.
- **Roles.** Here, the common project and/or program roles are outlined with respect to the project.
- **Business objectives.** The project purpose and objectives are described in further detail, including qualitative or quantitative metrics where possible. The language should reflect the business value of the project, its key deliverables, and desired outcomes.
- **Business case summary.** A concise description of the business justification for the project.
- **Project assumptions.** A statement of the key assumptions that have been used to inform the project schedule, scope, and budget estimates.
- **Project guidelines.** An outline of the decision priority that project managers should consider in order to empower project managers to make decisions. This is especially valuable when the project sponsor is not available to make decisions.
- **Project risks.** The key risks (opportunities or threats) that could confront the project, and how.
- **Project resources.** A description of the vital and scarce resources required for the implementation of a project. This includes the particular skills, technology tools, environment, and other potential needs.

- **Desired deliverables or project results.** The desired deliverables or outcomes of the project. For each deliverable or outcome, there is a concise description of the expected quality standard and its priority, for example, 1=*Must Have*, 2=*Should Have*, and 3=*Would Like To Have*.
- **Planned approach.** A description of the high-level approach to completing the project, including steps to jump-start the project. It is implicitly understood that situations and changes in the project may change the approach, without having to edit the charter.

Tools: This book includes the following templates for considerations. These tools are in Appendix A. Specifically:

- Template 2: Project Charter on page 265
- Template 7: Risk Register on page 274
- Template 8: Issue Log on page 275
- Template 9: Project Assumption Log on page 275

Insights from the front line: The author of this book, Dr. Wu, created a comprehensive charter for one of the major projects that he managed for Montclair State University. Click the link here to view the Canvas Learning Management System Implementation Project Charter: www.montclair.edu/media/montclairedu/oit/canvasmedia/Canvas-Project-Charter.pdf.

The completion of the project charter, whether it is accepted or rejected, formally ends the Project Initiation Phase. With the approval, the project then enters the next major phase – Preparation, which is the topic of the next chapter.

4.6 Project Management in Motion

This book is intended for a wide range of audiences and across various industries and functions. Therefore, to balance between competing demands of ease of use, comprehensiveness, modularity, flexibility, and upgradeability, the author has adopted a modular format to this book. This includes the relatively short chapters on key topics. In addition, the appendices include three vital sections:

- Appendix A contains a list of commonly used project management templates for both predictive and adaptive project management approaches.
- Appendix B contains an integrated case study based on a fictitious global company undertaking a number of projects and confronting various project challenges. The case maps closely to the chapters and sections, key concepts,

and relevant tools of this book. The goal is to provide instructors, students, and practitioners with a realistic project example to practice and apply the key learnings in the book. For more information and additional case studies in which the author plans to create over time, including potential collaboration opportunities, visit www.optimizepm.com.

- Appendix C contains a glossary of selective terms, reprinting with permission from Dr. Te Wu and Mr. Brian Williamson's book titled "*The Sensible Guide to Key Terminologies in Project Management*", iExperi Press, Montclair, NJ.

Chapter 5

Preparation – Planning to Achieve Optimal Implementation

Summary

After reaffirming the necessity of projects, typically by accepting the project charter by both the project sponsor and project manager, the project now shifts from the Initiation Phase to the Preparation Phase. The purpose of the Preparation Phase is simple – drive as much clarity as possible and prepare project teams before organizations commit to significant resources required for project implementation. This is achieved through the appropriate level of planning for the project.

The extent and level of clarity differs depending on the approach, the organization's culture, and methodology. On methodology, the level of planning can vary greatly among the various methods of project implementation. In perhaps overly idealist terms, the predictive project management approach requires detailed level of specificity before committing to action. Understanding, allocating, and assigning exactly who is doing what and at what time and place are some hallmarks of the predictive approach. Naturally, it is difficult for most complex and large projects to achieve this degree of specificity, but the intent is driving toward details.

In adaptive methods, the specific clarity often matters less than the overall construct of how the project team will pursue the project by analyzing factors such as the feasibility of modular implementation, the number of iterations or sprints, the duration of each iteration, customer review, and the dedicated core team responsible to drive the project's completion. The Preparation Phase can be more directional with a more thorough plan only for the next iteration or rolling wave. This way, detailed planning and preparation occur repeatedly at the beginning of each iteration as details of future iterations emerge. Hybrid approaches tend to pursue an approach that balances the need for clarity with demonstrating results early and getting customer feedback quickly.

This chapter addresses these three fundamental questions:

1. What are the key questions to address during project preparation?
2. What is project complex and how to manage them?
3. How to work with people – acquiring, building, mobilizing, managing, and leading teams?

5.1 About Preparation Phase

The Preparation Phase is generally the most intense period for project professionals to apply the technical and conceptual skills associated with the knowledge domains. This phase often starts with the formal acceptance of the project charter and ends with the creation of the project management plan (or equivalent). Depending on the implementation approach, the level of clarity and accuracy required, as exit criteria to this phase vary greatly. In traditional, predictive project management approach, the requirement for specificity is much higher than for the iterative or Agile approach. But at the minimum, the common exit criterion is that project managers are prepared for implementation and project teams are mobilized.

There are many important questions that project managers should address, regardless of the implementation approach. Table 5.1 captures some key questions during the Preparation Phase.

The answers to these and other questions are addressed throughout this book and especially in Part 3: Knowledge Domains. Here in the next section, this book will address the challenge of project complexity.

5.2 Project Complexity

On simpler projects, the power of project management is often muted. Occasionally, people may question why the application of project management, as project management itself is an overhead and requires precious resources. This is one reason that the author does not believe that project management should be applied on all projects. But on larger and complex projects, the importance of project management becomes self-evident. Complexity is relative; what's complex to one organization may be simple for another; and therefore, complexity can be difficult to define. According to Navigating Complexity (2014), complexity is "a characteristic of a program or project or its environment, which is difficult to manage due to human behavior, system behavior, and ambiguity" (Table 5.2).[1]

---

[1] Project Management Institute, 2014. *Navigating Complexity: A Practice Guide*, Project Management Institute, Newtown Square, PA.

Table 5.1 Project Preparation Questions

#	*Knowledge Domain*	*Key Questions during Preparation*
1	Integration management	How to manage the dependencies? How to manage the potential gap in responsibilities and communication between various project leads? How to tackle project changes?
2	Stakeholders management	How to best engage them? Which stakeholder's support is essential for the success of the project? How to "convert" blockers to supports?
3	Scope management	What are the detailed requirements? Are there situations or conditions in which these requirements change? What are the task packages? Does the project outcome and deliverable have long-term impact, such as changes to business operation?
4	Schedule management	Are there task sequencing considerations? What is the more specific task-level start and completion date and time? What are the specific dependencies?
5	Resources management	What are the detailed resource needs? Are there substitutional resources, in case we cannot find the optimal resources? For people, should we onboard internal employees? What if the skills are not an exact fit? Should we train people internally? What are the implications of team member location? What is the composition of the team, by skills, by locations, by employees, consultants, and vendors? What do they want from the project? How to best reward and motivate them?
6	Cost management	Are there additional concerns such as foreign currency exchange rate for global projects, time value of money for long duration project, or funding restriction on complex funding arrangements?
7	Communication management	Who needs to know what and when? How to best communicate with the key stakeholders?

(Continued)

Table 5.1 (*Continued*) Project Preparation Questions

#	*Knowledge Domain*	*Key Questions during Preparation*
8	Risk management	Who are the risk owners? How to develop risk response plan? For selective risks, do we need contingency plans and risk mitigation plans? How do we monitor risks?
9	Quality management	What are the specific functional and use considerations? How to ensure the project deliver the right quality?
10	Supply chain management	What should be the process of acquiring external resources? For example, should we implement one or more of the following processes: Request for Information, Request for Quote, and Request for Proposal?
11	Conflict management	How to deal with these conflicts effectively and prevent them from vicious spiraling? What are the best techniques and when to apply them?
12	Governance management	What should be the structure of the governance team? What are the roles and responsibilities managing the governance processes? Are there thresholds for these responsibilities?
13	Issue management	What are the root causes of these issues? How to effective address the issues to minimize their impact and potentially prevent future ones from occurring?
14	Adoption management	What is the organization's change readiness? Will the organization embrace the projects, as a part of project implementation and as a part of the ongoing operation of the project deliverables? Who will be impacted and to what extent? Should we develop plans to ease the change and minimize the impact? What does success look like from an organization and people perspective?
15	Operation management	Assuming there are operational impacts, what are these impacts analyzed at a detailed level? How to plan for these impacts? Who and how will these impacts be managed? When should the project start planning for the operational impact?

Table 5.2 Complexity Factors

Complexity Factor	*Sources of Complexity*
Human behavior	• Interplay between power, politics, personalities, and relationship • Multiple levels: individual, project team, organization, and industry
System behavior	• System connectedness increases the challenge of integration • Dependency management is crucial to achieve an order of implementation • System dynamics can be unpredictable, especially when integrating new systems
Ambiguity	• Uncertainty comes from unpredictable situations • Emergence is the unanticipated change that arises out of unknown, unanticipated, or deemed unimportant sources of change

5.2.1 Examples of Project Complexity – Problems and Potential Solutions

To deal with complexity, project managers must prepare the project through a detailed analysis and understanding of the project environment. Below are selective examples of complexity in projects:

- Environmental ambiguity
  - Highly competitive business environment resulting uncertainties from rapidly developing situations
  - Fast-moving technology development, for example, in hybrid information technology, artificial intelligence, and quantum computing
- System behavior
  - Depending on a range of factors, some of which not controlled by the project team, project managers must adopt an implementation approach among multiple competing methodologies such as waterfall, Agile, Kanban, DSDM, or some version of hybrid
  - On larger projects or in a multiproject environment, there are often multiple suitable options (or no good option)
- Human behavior
  - Working with difficult personalities or customers who cannot make up their minds
  - Transitioning in organization leadership, adjustments in scope and strategy, changing resources needs, and shifting priorities

Table 5.3 highlights selective common complexity challenges and good practices to tackle them.

Table 5.3 Complex Problems and Solutions

#	*Knowledge Domain*	*Problems*	*Potential Solutions*
1	Integration management	Managing change and distilling the impact of change to all areas of the project are perhaps two of the most difficult but common tasks in complex projects.	Apply a rigorous project management approach, especially with strong governance to enforce strong execution.
2	Stakeholders management	Complex projects often have high numbers of stakeholders and many with diverse interests. Also, the human behavior can be unpredictable.	While partial attention may suffice on smaller projects, full engagement is required on complex projects.
3	Scope management	In a complex environment, it's near impossible to define project scope precisely at the beginning – either the requirements are unknown or clients are changing their minds.	Consider an Agile approach that maximizes flexibility but also with diligent planning.
4	Schedule management	On complex projects, determining dependencies, sequencing activities, and estimating durations that are challenging and susceptible to change.	Work with the best experts, and adopt a standard process for managing schedule.
5	Resources management	When dealing with truly complex projects, it may be extremely difficult to find the appropriate resources – people and nonpeople. In some cases, resources can even be difficult to identify.	Be flexible and prepared to work with alternate resources and methods.

(*Continued*)

Table 5.3 (*Continued*) Complex Problems and Solutions

#	*Knowledge Domain*	*Problems*	*Potential Solutions*
6	Cost management	Estimation of costs accurately, especially in the beginning of the project, may be impossible.	Set the expectation with the sponsors and customers, and be transparent about the challenges and reality. Do not avoid difficult conversations.
7	Communication management	This is often the #1 cited reason for project failure. On large and complex projects, there are usually many people with communication channels increasing geometrically.	Employ appropriate communication methods, and manage meetings effectively.
8	Risk management	The combination of risk and complexity stands out as one of the most important practices where negative risks (threats) and complexity can be a toxic mixture and the positive risks can bring opportunities.	Apply appropriate risk management practices diligently.
9	Quality management	Quality can be fluid, since the project is difficult to define. Furthermore, there can be diverse opinions requiring negotiation.	Seek common ground, and enforce decisions. Establish checkpoints to enable greater flexibility and also transparency.

(*Continued*)

Table 5.3 (*Continued*) Complex Problems and Solutions

#	*Knowledge Domain*	*Problems*	*Potential Solutions*
10	Supply chain management	Finding the optimal supplier and shaping contracts advantageous to the project can be exceedingly challenging, especially when there is often a lack of negotiating power.	Be prepared to be flexible, with methods, contract structure, and resources.
11	Conflict management	Intractable disagreements compounded by multiple factors such as a clash of priority, divergence on process, and interpersonal or interorganizational discord.	Attempt to find common grounds among competing parties and untangle the intricate mess.
12	Governance management	Agreed decisions fail to stick or decisions are not widely supported by the organization.	Create a governance structure and processes that provide legitimacy and authenticity.
13	Issue management	Incidents that are difficult to detect with no clear root causes or obvious solutions are inherently difficult to tackle.	Assemble a diverse team of experts who can examine beyond the obvious to determine solutions.
14	Adoption management	Organization changes that emotional stress in the recipients. When emotions are involved, situations become more convoluted.	Retain organizational change experts to get ahead of noises and imagination through clear communication and trust building exercises.
15	Operation management	Severe impact to the operational environment but with little forethoughts or planning.	Start planning for the "day after" the project completion early in the project and include operational experts in the project team.

5.3 Identifying and Confronting Project Complexity

Complexities on projects can be hidden, and they can slowly creep into the project. By the time project professionals recognize the complexity, there are often less time to manage them properly. Therefore, it is important for project professionals to be proactive and look for concerns that may result in complex situations. Below are eight common warnings of complex situations:

1. **Large matrixed involvement** with many stakeholders and teams. This can result in in many interlocking requirements across cross multiple teams and functional areas.
2. **Deliverables are ambiguous** and perhaps unreasonably vast. Worse, they may have many strong dependencies.
3. **Conflicting or poorly aligned objectives** possibly as a result of sponsorship confusion. This can lead to a situation where there are too many executives in charge, but no one owns the entire project.
4. **Technology challenging** especially when working with leading edge and occasionally bleeding edge technology.
5. **Highly stressed teams resulting in departures** of important project team members. This can be a special concern on Agile teams, as they tend to work more intensely and with team members often playing multiple roles.
6. Challenging and perhaps unreasonable **timeline, resource, and/or budget constraints** creating unsustainable constraints and stress for the project team.
7. **High volume of change requests** requiring constant changes to the projects. In some extreme cases, the cost associated with change may be higher than the original project.
8. **Multilayer project/program governance** structures often involving with one or more C-level executives. This significantly slows the decision-making process, while the project continues to burn.

By the very nature of a complex situation, there is no one perfect way to address project complexity. Project managers need to develop a strong set of technical, behavioral, and people skills to manage complex situations proficiently. Below are six good project manager attributes when dealing with complex projects:

1. **Adaptable.** Must adapt the approach to the environment of the project/program. Consider with an open mind the optimal approach, e.g., waterfall, Agile, and hybrid.
2. **Collaborate.** Work closely with core team members, and communicate often. Build trust with important stakeholders.

3. **Proactive.** Think ahead and look beyond the next bend in the project. Be predictive of future activities, risks, and issues, and start navigating toward optimal engagement.
4. **Self-aware.** Understand one's strengths and weaknesses, and be willing to leverage expertise and experience of others within the project team. Delegate as appropriate, especially be willing to ask sponsors for help.
5. **Confront.** Engage challenges early, and head on to find collaborative ways to solve problems.
6. **Lead.** Be willing to take chances, make difficult choices, motivate and influence stakeholders, resolve conflicts, negotiate solutions, and solve problems.

For a discussion of additional planning processes, refer to Part 3: Knowledge Domain for a detailed examination of them for each of the twelve knowledge domains.

5.4 Project Management in Motion

This book is intended for a wide range of audiences and across various industries and functions. Therefore, to balance between competing demands of ease of use, comprehensiveness, modularity, flexibility, and upgradeability, the author has adopted a modular format to this book. This includes the relatively short chapters on key topics. In addition, the appendices include three vital sections:

- Appendix A contains a list of commonly used project management templates for both predictive and adaptive project management approaches.
- Appendix B contains an integrated case study based on a fictitious global company undertaking a number of projects and confronting various project challenges. The case maps closely to the chapters and sections, key concepts, and relevant tools of this book. The goal is to provide instructors, students, and practitioners with a realistic project example to practice and apply the key learnings in the book. For more information and additional case studies in which the author plans to create over time, including potential collaboration opportunities, visit www.optimizepm.com.
- Appendix C contains a glossary of selective terms, reprinting with permission from Dr. Te Wu and Mr. Brian Williamson's book titled "*The Sensible Guide to Key Terminologies in Project Management*", iExperi Press, Montclair, NJ.

Chapter 6

Implementing Projects – Getting It Done

Summary

Einstein observed "genius is 1% inspiration and 99% perspiration". This analogy applies well to projects. On complex projects, for all the intense planning and preparation, the real journey starts with project implementation. The Project Implementation Phase is generally the most expensive and most resource intensive and has the longest duration of all the phases. Rarely do project plans survive the major early challenges without some casualty of assumptions that require change.

As a part of project implementation, project managers spend a considerable amount of time monitoring the progress of their projects, comparing the actual with the planned, and making adjustments to control the process and the outcome. Controlling is considered a primary function of management because most complex endeavors require active steering and adjustment during the journey.

For project managers, this is both exciting and daunting. It is exciting because now the planning is finally in motion and project execution starts. It is daunting because even with the best of planning, problems are inevitable, new risks are likely to surface, and changes will probably occur.

This chapter covers five important questions on implementing projects:

1. What are some common project implementation challenges and how to address them?
2. Why is monitoring and controlling so crucial to project success?
3. How to control projects effectively?
4. How to be a better project manager?
5. How to deal with disagreements and conflicts?

Stakeholders	Success Factors	Success Criteria
	Project Success	
Executive leadership Client (Internal / external)	Success Factors • Organizational structures • Management practices and values • Executive support • Clear vision and scope • Adequate resources' allocation • User involvement • Project management success	Success Criteria • Alignment of project's output with organizational strategy • Organization's benefits realization
	Project Management Success	
Project Leaders & Teams	Success Factors: • ***Project Planning*** • ***Project Execution*** • ***Project Reporting*** • Project team competencies	Success Criteria • Quality -Met the original goals/scope • Cost- Finished within the budgets • Time- Finished within the scheduled times

Figure 6.1 Project success and project management success. (Training and consulting content from PMO Advisory LLC. Reprinted with permission.)

6.1 The Art of Execution

The execution of project work takes place primarily in this phase of the project life cycle. While these processes, and their inputs, tools, and outputs, provide an orderly and scientific approach, project management is not always this orderly. The project management discipline is not always as easily organized, examined, dissected, measured, and reconfigured. While there have been many advances toward best practices for achieving project success, the failure rate of projects and their management continues to be disappointingly high.

Research has found that project success rates have increased by only 10% over the past 21 years.[1] Nearly 50% of projects have been found to exceed their original schedules and experience scope creep or significant scope changes.[2] Many also exceed their initial project budgets. A recent PMI study found that $99 million of every $1 billion dollars spent on projects is wasted.[3] Furthermore, significant gaps are often found between project success and project management success, which are characterized by different stakeholders, success factors, and success criteria (see Figure 6.1).

---

[1] Johnson, J. & Mulder, H. 2016. CHAOS chronicles, focusing on failures and possible improvements in IT projects. *Conference Paper. The 10th International Multi-Conference on Society, Cybernetics and Informatics: IMSCI 2016*, Orlando, Florida, USA.

[2] Project Management Institute. 2014. *Pulse of the profession in-depth report: Enabling organizational change through strategic initiatives*. Newtown Square, PA.

[3] Project Management Institute. 2018. *Pulse of the Profession: Success in Disruptive Times*. Newtown Square, PA. P2.

Total Project Success is the achievement of both project management success and project (or product) success. For organizations sponsoring the project, the project success may be even more encompassing than building the product. For example, when the Apple iPad® was first released on April 3, 2010, the First-Generation iPad took the world by storm. A sleek piece of technology weighs little more than a pound, but with the power to access over 100,000 applications and practically all the Internet. Rumors surfaced that Hewlett Packard was also ready with a tablet of their own but decided to postpone the introduction. Assuming that Hewlett Packard developed a quality tablet on time and within schedule, the project would have been deemed successful. And yet, when the HP product compared with the Apple's iPad, the company went back to the drawing board. The business value that Hewlett Packard hoped to achieve was not realized. This means that for the company and its shareholders, the project was at least a partial failure.

6.2 Monitoring Projects

Martin Cobb once said that "*We know why projects fail; we know how to prevent their failure – so why do they still fail?*" Throughout the project, the project manager is responsible for monitoring and controlling project work. This is the process of monitoring, analyzing, and reporting the overall project progress. In general, the project manager needs to monitor time (schedule), cost (budget), and scope (project performance). However, there can be additional considerations, such as quality, vendor performance, adoption, and team satisfaction, along with sponsor and executive support. Due to time constraints, it is important to focus on collecting only necessary information and to avoid collecting too much information. The next step is to design a reporting system that distributes the relevant data to the right people in a timely and understandable manner.

Definition. Common activities performed in monitoring and controlling projects include the following:

- *Monitoring.* This involves collecting, recording, and reporting project performance information.
- *Evaluating.* Based on data collected by monitoring, judgments are made regarding the quality and effectiveness of project performance and deviations from the plan.
- *Controlling.* When deviations occur, actual performance is brought back into compliance with the plan.
- *Auditing.* This provides objective analyzes and reviews of the project and project management processes to ensure performance.

All stakeholders should be tied to the reporting system, and reports should address each level at the appropriate depth and frequency. Lower levels of management require detailed information, whereas senior management levels usually only require overview reports. The frequency of reports is typically higher at low levels of management and less frequent at higher levels of management. Reports should contain relevant data that are made available in time to control, in the event of issues. Mysterious stakeholders, unconstrained constraints, suspicious status reports, discord and drama, and unresponsive sponsors can be signs of issues.

The distribution of project reports is dependent on their content. *Routine Reports* are issued on a regular basis, or each time the project reaches a milestone. *Exception Reports* are generated when an exceptional condition occurs, or as an informational vehicle when an unusual decision is made. *Special Analysis Reports* result from specific studies commissioned to look into unexpected problems. The benefits of detailed and timely reports are a mutual understanding of the project goals, awareness of the progress of parallel activities, understanding the relationships between tasks, early warning signs of problems, minimized confusion, higher visibility of top management, and keeping the client up to date.

6.3 Managing Special Situations in Projects

In an ideal world, projects proceed smoothly from the Preparation Phase, with no deviations between actual and targeted outcomes. However, this scenario is unfortunately rare. Projects are frequently confronted with problems and challenging situations. Assumptions, risks, issues, or changes can cause these special situations. An *assumption* is an aspect of the planning process considered being certain, real, or true, without demonstration or proof.[4] They affect all aspects of project planning and are part of the progressive elaboration of the project. Project teams frequently identify, document, and validate assumptions as part of their planning processes. However, while assumptions are generally informed, they do also involve a degree of risk, and they can change as more information becomes available. Consequently, it is important to manage assumptions as risks (or issues if changes occur).

Risks are potentialities that, if they materialize, can have impacts on one or multiple objectives in a negative or positive manner, in the form of resources, performance, quality, or timeline. *Changes* are alterations that result in deviations from previously agreed-upon standards. These can be related to schedule, resources, cost, scope, quality, project management processes, or other components of the project. Examples of each risk, issue and change are shown in Table 6.1.

[4] Williamson, B. & Wu, T. 2019. *The Sensible Guide to Key Terminologies in Project Management*, iExperi Press, Montclair, NJ. Glossary.

Table 6.1 Risks, Issues, and Change

Type	*Risk*	*Issues*	*Change*
Example	A business department may not provide feedback in time.	A business department did not provide feedback by the deadline, which was a few days ago.	The developer agreed to a new feature without informing the project manager.

The purpose of *risk management* is to reduce overall project risk to a level that is acceptable. The cost of risk mitigation is often higher when risks are discovered later on in the project. Best practices call for the creation of a risk register, which contains the risk description, quantitative or qualitative analysis of risk, the probability of occurrence, the severity of impact, the risk owner, a risk response or mitigation plan, and contingency plan.

The goal of *issue management* is to tackle or resolve problems successfully, remove obstacles, or address real concerns. Unlike risks, the probability of an issue is 100%, because it has already occurred or is currently occurring. Similar to risk, managing an issue requires an issue log. This is a living document that lists all project issues, including an analysis of urgency and importance, and a course of action (sometimes immediate) for addressing issues. As issues are identified, they should be captured and documented. Despite being the owners of project issues, project managers should assign issues to trusted team members.

At its core, *change management* is about making change explicit and known. The decision to accept or deny change requires active participation and acknowledgment from all relevant stakeholders. Unknown changes result in scope creep, cost overruns, schedule delays, quality problems, and overworked staff. In some situations, changes are made without the project manager's agreement or knowledge. Similar to risks and issues, changes should also be formally captured in a change registry.

6.4 Controlling Projects

Project managers are responsible for the monitoring and controlling of the entire project, across all ten knowledge domains. Control is considered a primary management function, because large projects rarely (if ever) execute in accordance with the plan without adjustments. Control is the process of making adjustments to enable projects to remain on course. Project managers are the stewards of organizational assets and resources and have fiduciary responsibilities to control human resources, physical assets, and financial resources. Project managers are expected to motivate people and enable them to perform satisfactorily, to conduct performance evaluations, recognitions, and to award other rewards (or penalties) to control performance. Project managers are also responsible for controlling the use and access of physical assets and ensuring the necessary preventative and corrective maintenance of equipment and systems. Project managers must also manage project budget, including cash flow and capital investment, and must exercise careful diligence, especially in a regulatory environment.

Control can take form in many ways and across many dimensions:

- **Preventative controls versus repairs.** *Preventative* controls seek to address future problems before they occur. Generally, these work in conjunction with risk management to anticipate future challenges. *Repairs* are designed to address problems after they occur and improve the quality of subsequent endeavors.
- **One-dimensional versus multidimensional controls.**[5] *One-dimensional* project control refers to managing specific issues, such as scope creep. *Multidimensional* project control refers to the integrated management of several dimensions, such as scope creep that results in cost, resource, and schedule increases.
- **Go versus no-go control**. With this type of control, progress is determined by sufficing previously agreed performance metrics. This requires careful management of data collection, analysis, and reporting.
- **Phase gate (also known as stage gate)**. Control is exercised by requiring that projects meet specific criteria before advancing to the next phase or stage.

earned value management (EVM) system is a project management technique for measuring and controlling project performance objectively. The technique is simple, consisting of only a few essential concepts, such as "*value*" (the worth of project progress and accomplishment), "*planned*" (targeted goals and objectives), and "*actual*" (what has been achieved). EVM primarily measures two dimensions, namely cost and time.[6] Using basic arithmetic, EVM provides a robust set of project management performance measures and calculations. Two of the most popularly used indices include the Cost Performance Index and the Schedule Performance Index.

For a more complete discussion of earned value and a detailed example, please refer to Chapter 13 on Cost Management in which a section focuses on EVM.

6.5 Project Challenges during Implementation

Projects can fail for a wide variety of reasons. Some of the most common reasons are poor project sponsorship or leadership, unreasonable expectations, insufficient resources or time, unclear project expectations or sponsorship, ambiguities and

---

[5] Rozenes, S., Vitner, G., & Spraggett, S. 2006. Project control: Literature review. *Project Management Journal*, 37(4): 5–14.

[6] Fleming, Q. W., Koppelman, J. M., 2016. *Earned Value Project Management*. Project Management Institute.

changes to agreed-upon requirements, inadequate project management, poor planning, confusing methodology, and a vicious combination of issues, risks, and conflicts. These challenges can present themselves in different ways and during different phases of the project.

Challenges during the *Ideation Phase* include defining the project and its high-level requirements and making a case for its implementation. *Initiation Phase* includes solidifying project scope, selecting a capable project sponsor, project manager and core project team, agreeing on approach and methodology, agreeing on a baseline for project success, and developing reasonable project estimations. *Preparation Phase* challenges arise when identifying requirements, creating solutions to achieve requirements, developing a thorough schedule with agreement from the project team, acquiring key resources, and obtaining approval for the Project Management Plan. Challenges in the *Implementation Phase* are typically the management of people (especially the most irreplaceable resources), engaging stakeholders and managing their expectations, managing organizational politics and personalities, making trade-off decisions (especially when there is imperfect information available), anticipating the unexpected, and effectively navigating risks, and performing integrated change management activities. Challenges in the *Transition and/or Closure Phase* involve getting people to adopt the project deliverables, fixing defects and addressing punch list items to achieve customer satisfaction, transitioning project work to operational teams (and obtaining their sign-off), achieving agreements with vendors to address punch list items, updating important project documents before the team demobilizes, and obtaining final client sign-off to close this phase of the project.

Fortunately, these challenges can be addressed through project management best practices, such as thorough and collaborative planning. Even though plans change, the exercise of engaging relevant stakeholders in collaborative planning upfront is beneficial when confronting changes later on. Another project management best practice is to manage stakeholders well, particularly on complex projects where the definition of project success can be elusive. Stakeholders will be critical in achieving success and must thus be engaged and onboard with key decisions. Furthermore, project managers should avoid uninformed assumptions about key project activities and decisions. All assumptions should be communicated, checked, audited, and verified. When working with new teams or vendors, project managers should always validate important information. Issues and risks should be tackled head on, because trying to avoid issues and risks will likely only delay the damage. Finally, effective project managers need the support of important stakeholders, and to gain this support, they must build trust. Part of building trust is to communicate bad news early on and deal with it appropriately.

6.6 Think Like a Project Manager

Projects are challenging and require strong leadership and management to achieve their intended success. This is why the project manager plays such a critical role. To be an effective project manager, it is important to start by thinking like a

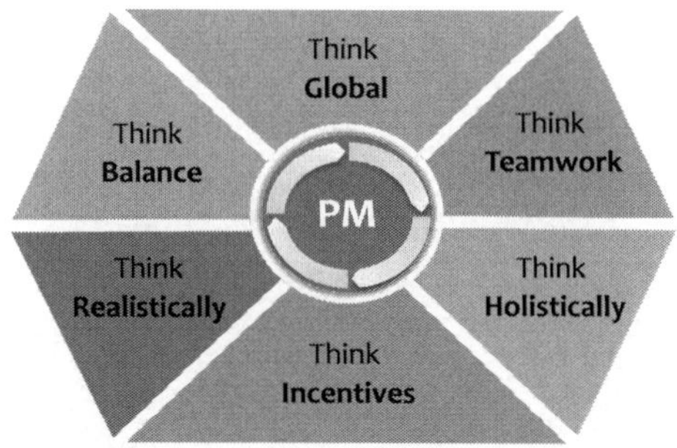

Figure 6.2 How to think like a project manager. (Training and consulting content from PMO Advisory LLC. Reprinted with permission.)

project manager. Six ways of thinking have shown to be effective in today's organization: think *balance*, think *global*, think *teamwork*, think *incentives*, think *realistically*, and think *holistically*. Figure 6.2 illustrates the six ways of thinking:

To think about *balance*, project managers must mentally balance optimism, skepticism, and the occasional cynicism. They must be able to balance analysis with action in order to avoid analysis paralysis and/or reckless reactions. This means that project managers think continuously in terms of trade-offs. Project managers are also culturally sensitive, respectful, and open-minded in their thinking, because they embrace the *global* diversity of people, their differences, culture, and ideas on their projects. Project managers think in a way that leads them to take ownership and take charge of their *teams* and to set an example for others. They are willing to challenge their team, while also being willing to yield expertise themselves. Thinking in terms of *incentives* is a key approach for project managers to drive their team. All people are driven by incentives, both tangible (such as compensation) and intangible (such as recognition, respect from peers, and a sense of well-being). Through this thinking, project managers are able to design metrics that reinforce the right incentives carefully. Project managers must be fiercely *realistic* and practical in their thinking around people. This thinking will lead to more realistic expectations and estimates. Finally, project managers must at all times maintain the strategic perspective of the bigger project goals in all of their thinking. *Holistic* thinking recognizes that a project is often greater than a set of deliverables and may have interdependencies with other areas.

6.7 Achieve Sound Decisions

The cost of poor decision-making in project management can be high and sometimes irreversibly damaging. The evidence of this has been vast in business and history. The Samsung Galaxy Note 7 that exploded in 2016 was banned from flights and resulted in a $5.3 billion loss. Samsung revealed that a rush to release the updated version contributed toward the fault.[7] Research in Motion (now Blackberry) achieved a 20% smartphone market share in 2010 and now holds less than 0.2%. AOL and Time Warner underwent the greatest merger and acquisition failure in recent history in 2000, resulting in a $100 billion loss in value. These examples demonstrate the cost of poor project decisions, but they also point toward the benefits of sound decision-making. Companies such as Apple (the most valuable firm by market share in the world), Facebook (which achieved 1 million active users in just 10 months and 1 billion active users in 8 years), and PMI (the PMP certificate grew from about 7,000 active users in 2003 to over 960,000 active PMPs in 2019) have managed to achieve significant growth and revenue through sound decision-making.

However, making sound decisions is not always simple or easy to do. There are a number of challenges that project managers encounter when making decisions. Possibly the most common occurring of these is insufficient, incorrect, or incomplete information available to make a timely, informed decision. In these situations, it becomes tempting to fill in the blanks with "gut feel", which exposes decision-making to emotion and bias. This can often lead to poor decisions. However, it is still important to acknowledge feeling, passion, and values in decision-making, as long as it does not cloud judgment. This means trusting one's intuition and experience, without discarding an objectivity based on facts and data.

Another common challenge to making good decisions is the time pressure of projects. Too much emphasis on speed and agility sometimes results in rushed decision-making, where some of the most important information has not been properly considered. However, taking too long to react can also result in analysis paralysis and negatively impact the project. Thus, project managers must find a balanced approach where they are able to deal with some level of uncertainty, or accept that there is more information than time to work through it, and arrive at the best possible decision given the circumstances. Finally, decisions are challenged by their complexity and the multiple, changing factors that influence them. This often results in decisions being second-guessed or backtracked and key decision-makers changing their minds. However, if decisions are not observed and respected, then even the best decisions will not lead to positive outcomes.

In project management, the primary goal is often *not* to make the "best" decision, but to adopt a decision-making method that is justifiable through its legitimacy and consistency. Arguably one of the most effective methods for addressing the challenges of decision-making, and providing legitimacy, is the Analytical Hierarchy Process, which was discussed in Chapter 3.

---

[7] Kovach, S., 2017. *Samsung Announces What Caused the Galaxy Note 7 to Overheat and Explode*. Business Insider, Retrieved from www.businessinsider.com/samsung-issues-galaxy-note-7-battery-report-2017-1.

6.8 Project Management in Motion

This book is intended for a wide range of audiences and across various industries and functions. Therefore, to balance between competing demands of ease of use, comprehensiveness, modularity, flexibility, and upgradeability, the author has adopted a modular format to this book. This includes the relatively short chapters on key topics. In addition, the appendices include three vital sections:

- Appendix A contains a list of commonly used project management templates for both predictive and adaptive project management approaches.
- Appendix B contains an integrated case study based on a fictitious global company undertaking a number of projects and confronting various project challenges. The case maps closely to the chapters and sections, key concepts, and relevant tools of this book. The goal is to provide instructors, students, and practitioners with a realistic project example to practice and apply the key learnings in the book. For more information and additional case studies in which the author plans to create over time, including potential collaboration opportunities, visit www.optimizepm.com.
- Appendix C contains a glossary of selective terms, reprinting with permission from Dr. Te Wu and Mr. Brian Williamson's book titled "*The Sensible Guide to Key Terminologies in Project Management*", iExperi Press, Montclair, NJ.

Chapter 7

Project Transition and/or Closure – Celebrating Success

Summary

As a temporary endeavor, by definition, all projects end. Some projects end happily, whereas others end less so. But regardless of how projects end, project managers are responsible for the proper closing of projects and whenever necessary, transitioning the deliverables to operation. Project Transition and/or Closure Phase can be a busy time for project managers, as they wind down the project and transition the deliverable to another project or to operation.

This chapter explains the closure and transition activities including finishing the "punch list" of tasks, obtaining sponsor or customer sign-off, demobilizing the project team and reallocating resources, paying project vendors, conducting lessons learned, updating key project deliverables and artifacts, transferring project knowledge to operational teams, preparing for the final project report, and perhaps the most important activity – celebrating with the project team.

This chapter explains three important questions about project transition and/or closure:

1. Why project transition and closure are important?
2. What are the different types of project transition and closure?
3. What are some important challenges in project transition and closure and how to manage them?

7.1 Transitioning and/or Closing Projects or Phases

All projects and phases must come to an end. Ideally, this ending is brought about because the project has been successfully completed, objectives have been achieved, and stakeholders are satisfied. In this happy scenario, the project deliverable is fully completed with the client signed off on the work. Therefore, the current phase of the project is completed, and the project goals are achieved and the deliverables are integrated within the existing organization or transitioned to the next phase. In all of these scenarios, there is a transitioning from the project, its team members, knowledge and skills, systems and processes, and deliverables or project results, to become integrated into operations within the organization. Depending on the size of the project, planning activities for termination should start in the Preparation Phase of the project. Sometimes, a special transition manager is appointed to lead this activity. Today, a commonly used approach in implementing and closing projects is Benefits Realization Management. This is a collection of processes that ensures that the strategic benefits of projects are sustainable after project implementation is complete. This approach has been found to have a significant positive influence on project success.[1]

Insights from the front line. On large and complex projects, project success often means achieving all or nearly all the following:

- Mission has been achieved
- Business and technical requirements have been met
- Project schedule and plan have been met
- Project was completed on budget, or within preagreed tolerance ranges
- Top management support
- Client acceptance
- Team member satisfaction
- Vendor has successfully delivered results

Projects do not always end in successes and can have unhappy endings. Some projects are prematurely terminated, end abruptly, or limp along without ever coming to a closure, often resulting in wasted time and resources. Sometimes,

---

1 Serra, C. E., Kunc, M., 2015. Benefits Realization Management and its influence on project success and on the execution of business strategies. *International Journal of Project Management*, 33(1): 53–66.

this kind of ending is even necessary. This unhappy ending can occur for a number of reasons:

- Reprioritization is when the sponsoring organization shifts its priority and shuts down the project to conserve funding or resources, for new priorities. This is also known as *termination by convenience.*
- Sometimes, projects end abruptly due to poor project management – when the project depletes its funding, resources, or time without achieving the project objectives.
- This can also occur when a key vendor decides to pull out and the sponsoring organization decides to terminate the project instead of supporting it further.
- Projects that fade out and never come to a definite ending are commonly tagged as low-priority projects. They can also be "pet projects" that did not attain wide support. These pet projects were often not officially approved, or where the sponsor has left the organization and in turn left the project without sufficient organizational support. The project manager and/or team members may continue to work on the project, but likely at a reduced allocation as they become busy with other work.

Regardless of the reasons, all of these projects end without achieving their goals. Whenever possible, project managers on these projects should work professionally and diligently to salvage the knowledge and work achieved and shelve them for another day. As Woody Williams once said, "If you have never recommended canceling a project, you haven't been an effective project manager".

Regardless of how the project is terminated, one of the most common best practices in the Transition and/or Closure Phase is to conduct a *post project evaluation, also known as lessons learned.* The evaluation should contain information about what worked, what should be improved, lessons that should be repeated in the future, and lessons to be avoided. Project managers should also update and summarize the key project metrics and work with the team to update important documents, close vendor contracts and pay them accordingly, and reward and recognize key contributors. Finally, the project manager is responsible for ensuring the smooth transition of team members to other projects and operations, which can be difficult when the team has become close.

7.2 Specific Transitions and/or Closures

As the project comes to a close, there are specific transitioning and/or closure (dependent on the project) in all knowledge domains, including scope, Schedule, cost, quality, resources, communication, risk, supply chain, stakeholders, and integration. The approach must be tailored to the specific organization's policy and the project situation.

7.2.1 Scope Transition and/or Closure

Throughout the project, there are likely to be changes that are not fully incorporated in the project documentation. Thus, it is possible that the actual deliverables deviate from the plan. Thus, upon project closure, it is recommended that all scope-related deliverables that could be of use in the future should be updated so that they reflect the actual deliverables. This could include maintenance, training, minor enhancements, and major upgrades to the scope statement, scope management plan, requirements, traceability matrix, engineering artifacts, and technical specification.

7.2.2 Schedule Transition and/or Closure

Project managers should update the project plan to reflect the actual work performed within a given timeframe. The purpose of this is to provide input to lessons learned and create an auditable trail of work. Furthermore, if the project has multiple phases, the updated project schedule serves as an input for planning the next phase of the project.

7.2.3 Cost Transition and/or Closure

Given the sensitivity of finances, in most professionally run organizations, project managers must update the project budget to reflect actual spending, and analyze and explain when this breaches a preagreed threshold. From a project management perspective, overspending and underspending beyond a reasonable threshold and without sound explanation shows poor project management. The negativities pertaining to overspending are more obvious, since the project manager will need to request more funding. However, spending below the budget is also considered poor project management. This is because underspending indicates that the project budget was overestimated, and the unused funds will have a cost of capital. Unused funds are opportunity costs for the organization; by allocating those funds to this project, it also prevented the organization from investing in other project opportunities. Some important cost-related documents to update include the budget and cost analysis, cost management plan, and project cash flow or disbursement schedule.

7.2.4 Quality Transition and/or Closure

Project managers should build the project deliverables based on the agreed-upon quality plan in order to achieve deliverables that are both fit for function and fit for use. Specific documents include the quality management plan, acceptance criteria, quality assurance artifacts, test plans, defect log, or punch list.

The project manager should preserve the document that has been signed off by the customer, especially when the customers are external and the sign-off is required for payment.

7.2.5 Resource Transition and/or Closure

The project manager is responsible for the release, and sometimes the reassignment, of project resources. These can include people, such as employees reassigned to other projects, or consultants and vendors released from the project work. Resources can also include nonpeople assets, such as tools or materials that must be returned to their owners, or in some cases, maintenance work may be required to restore physical assets to preagreed conditions. Project managers should also make updates to the resource management plan, capability and capacity analysis, responsibility assignment matrix (RACI), performance evaluation of team members, and individual development plans for selective team members.

7.2.6 Communication Transition and/or Closure

In addition to documenting the effectiveness of communication efforts on the project in the lessons learned register, project managers should also update the project information management system and archive the documents in the system for future use. This includes creating metatags and other activities to enhance the reusability of project content. Furthermore, the communication management plan should be updated to reflect the actual plan. The lessons learned register is itself a form of communication that brings the project to a closure.

7.2.7 Risk Transition and/or Closure

The closure of risk will be largely dependent on the type of risk. Some risks are likely to outlast the project. Therefore, it is important for the project manager to transition the management of these risks to another member as required. There could be an organizational risk that does not materialize during project implementation, but after project closure, the risk materializes during the ongoing operation period or the next project phase. Therefore, it is important for the project manager to communicate with his or her counterpart in organizational operations and/or the next phase of the project. Specific artifacts to update for risk include the risk management plan, risk register, risk analysis, risk response plan, and lessons learned on risk implementation.

7.2.8 Procurement Transition and/or Closure

On projects that require external resources, project managers should work with the seller to review and properly close all contracts, such as statement of work, acceptance sign-off, performance reports, legal endeavors, such as dispute resolution or litigation, and financial activities, including making final payments. Project managers should also update procurement artifacts such as the procurement management plan, lessons learned, vendor performance evaluation, and related contracts and payment schedules. It may also be necessary to review and update the related issue log and risk registry.

7.2.9 Stakeholder Transition and/or Closure

Project managers should review the important stakeholder documents, such as stakeholder analysis, the stakeholder engagement plan, and stakeholder management plan, and make required updates in order for future projects to learn from this experience. Stakeholder updates should focus on situations, priorities, and roles, while avoiding the naming of specific individuals.

7.2.10 Project Integration Transition and/or Closure

The final integration checklist should include the following:

1. Ensure tasks are completed.
2. Close and update change requests.
3. Notify the customer of the closure.
4. Finish the paperwork – archive, file, index, and store.
5. Send out final invoices to the client (if applicable).
6. Redistribute resources.
7. Clear with legal counsel.
8. Determine what records to keep.
9. Assign support.
10. Close the project books.
11. Celebrate!

7.3 Project Management in Motion

This book is intended for a wide range of audiences and across various industries and functions. Therefore, to balance between competing demands of ease of use, comprehensiveness, modularity, flexibility, and upgradeability, the author has adopted a modular format to this book. This includes the relatively short chapters on key topics. In addition, the appendices include three vital sections:

- Appendix A contains a list of commonly used project management templates for both predictive and adaptive project management approaches.

- Appendix B contains an integrated case study based on a fictitious global company undertaking a number of projects and confronting various project challenges. The case maps closely to the chapters and sections, key concepts, and relevant tools of this book. The goal is to provide instructors, students, and practitioners with a realistic project example to practice and apply the key learnings in the book. For more information and additional case studies in which the author plans to create over time, including potential collaboration opportunities, visit www.optimizepm.com.
- Appendix C contains a glossary of selective terms, reprinting with permission from Dr. Te Wu and Mr. Brian Williamson's book titled "*The Sensible Guide to Key Terminologies in Project Management*", iExperi Press, Montclair, NJ.

3

KNOWLEDGE DOMAIN

Chapter 8

Project Integration Management – The Most Important Knowledge Domain

Summary

This chapter explains the important responsibility of integration management, including the development of the project charter and project management plan, directing and managing project work, controlling performance, and managing change. The critical processes of managing project knowledge and closing the project or phase are also discussed.

Even though project management is composed of over a dozen knowledge domains (and can be more depending on projects), projects are not managed discretely. A project manager cannot or at least should manage only the schedule and disregard scope, resources, or quality, for example. Integration management is the knowledge domain that ties all these together. Thus, sound decisions by project managers are likely be an optimized solution based on a holistic analysis of the many factors. Project managers are managers and leaders because they are responsible for looking beyond the details and make broader decisions that impact the entire project.

This chapter addresses these three fundamental questions:

1. Why is project integration considered to be the most important knowledge domain?
2. What is project integration?
3. How to plan, direct, and control projects more holistically?

8.1 Importance of Project Integration

While project managers are often able to delegate tasks and responsibilities to other leaders and managers, integration management is one area that project managers must own. Project integration management is the combination and synthesis of all project management knowledge domains and the associated processes required to achieve synergy and create a "whole" that is greater than the sum of the parts. Project integration includes activities such as resource allocation, balancing competing demands, weighing-up alternative approaches, adapting processes, and managing interdependencies. Project integration should be performed throughout the project life cycle.

On projects, the twelve knowledge domains of project management rarely work in isolation. Instead, the project manager must coordinate all knowledge and processes together in order to leverage complete project integration toward project success. For example, by adding more resources, the project schedule is often shortened, but costs will rise and quality may be compromised. Improved planning and communication between stakeholders assist in managing increased project risk, while carefully and strategically managing stakeholder expectations, and understanding the project requirements, can lead to increased resources and time.

EXAMPLE: THE CHALLENGE OF EXPEDITION

You are the project manager assigned to a client project to develop a new application. Before the project begins, you and your team map out two scenarios: (1) the Normal Condition, which is a realistic and manageable project schedule given the project requirements, and (2) the Expedited Condition, which is mapped on a shorter timeline. The client agrees to the Normal Condition, where the cost is $2,840 and the duration is 30 days. The project is scheduled to start in 30 days' time.

Five days later, the client contacts you with an urgent request. The marketing team has asked that the application be developed as quickly as possible, and the client is willing to pay 25% more to get this done. To manage the change, you need to take an integrated approach. You analyze the project activities and resource allocation and decide to crash the project. *Crashing* is a scheduling

			Normal		Expedited	
ID	Activity Name	Pre-requisite	Duration (Days)	Activity Cost	Duration (Days)	Activity Cost
A	Plan project		2	$ 240	2	$ 240
B	Develop requirements	A	4	$ 360	3	$ 600
C	Design solution	B	3	$ 300	2	$ 550
D	Develop codes	C	10	$ 750	8	$ 1 200
E	Testing	D	6	$ 540	4	$ 800
F	Training	E	4	$ 450	3	$ 600
G	Deployment	F	1	$ 200	1	$ 200
		Total:	**30**	**$ 2 840**	**23**	**$ 4 190**

Figure 8.1 Normal versus expedited schedule.

technique used to shorten the scheduled duration for the least incremental cost by adding resources. You have also considered *fast tracking* (another schedule compression technique where activities or phases normally done in sequence are performed in parallel for at least a portion of their duration). However, the risk is too high given that this is the first time the project team is working with the client, and the early experience suggests that caution is warranted.

Your next step is to develop an expedited model including a revised duration and expedited cost for the project (see Figure 8.1). You also invoke the integrated change management process (a process within the project integration management knowledge domain) to manage the change.

This change impacts two areas of the project: schedule and resources. To determine how many days the client would save by spending 25% more, you reevaluate the project plan and develop the following analysis (Figure 8.2):

			Normal		Expedited		Expedited Plan		
ID	Activity Name	Pre-requisite	Duration (Days)	Activity Cost	Duration (Days)	Activity Cost	Number of Days Saved	Total Expedited Cost	Cost Per Day
A	Plan project		2	$ 240	2	$ 240			
B	Develop requirements	A	4	$ 360	3	$ 600	1	$ 240	$ 240
C	Design solution	B	3	$ 300	2	$ 550	1	$ 250	$ 250
D	Develop codes	C	10	$ 750	8	$ 1,200	2	$ 450	$ 225
E	Testing	D	6	$ 540	4	$ 800	2	$ 260	$ 130
F	Training	E	4	$ 450	3	$ 600	1	$ 150	$ 150
G	Deployment	F	1	$ 200	1	$ 200			
		Total:	**30**	**$ 2,840**	**23**	**$ 4,190**			

Figure 8.2 Expedited plan.

To address the client's request, you recommend reducing the project by 4 days but incurring a cost increase of 23%. This would include a day saved on training ($150), 2 days saved on testing ($260), and 1 day saved on developing requirements ($240) for a total saving of $650.

However, after reviewing the recommendations, the client contacts you again. This time, the marketing team has revised its forecast and decided that the project must be delivered in just 20 days. The client is willing to "do what it takes" to complete the project within this timeframe, including increased expenditure and accepting more risk. The client has also offered to send three dedicated subject matter experts to colocate with the project team to enhance communication and problem-solving.

After further analyzing the project plan, you believe that the new deadline can be achieved by crashing the project and fast tracking simultaneously. You warn the client of the high risks, and the client insists that you proceed. Furthermore, the client promises a bonus of $1,000 beyond the expedited cost to motivate the project team. After receiving approval of the aggressive plan through the integrated change management process, you are ready to execute. The revised Gantt chart is shown in Figure 8.3.

By starting Activity E early, the project plan directs the testing team to focus on the codes that have already been developed. By starting Activity F, training, at the same time as quality testing, you plan to combine some testing activities with training. Testing will start while development is still in progress. The $1,000 bonus is also a strong motivator for your team. As this is the first project between these parties, both parties want a strong win.

Activity	1	2	3	4	5	6	7	8	9	10	11	12	13	14	15	16	17	18	19	20	21	22	23	24	25	26	27	28	29	30
A	A	A																												
B			B	B	B	B																								
C							C	C	C																					
D										D	D	D	D	D	D	D	D	D	D											
E																				E	E	E	E	E	E					
F																										F	F	F	F	
G																														G

Activity	1	2	3	4	5	6	7	8	9	10	11	12	13	14	15	16	17	18	19	20	21	22	23	24	25	26	27	28	29	30
A	A	A																												
B			B	B	B																									
C						C	C																		Fast Tracking					
D								D	D	D	D	D	D	D	D															
E														E	E	E	E								Crashing					
F																	F	F	F											
G																				G										

Figure 8.3 Original (above) versus expedited (below) project Gantt charts.

Today, project managers have a number of tools and practices available to assist with project integration management, from project management information systems (PMISs) and visual management tools, to knowledge management practices and hybrid approaches to project management. Using these tools and practices, project managers must adapt their integration approach to each project depending on its requirements. In Agile environments, a focus on collaborative and engaged decision-making during the project life cycle becomes particularly important for responsive integration management.

8.2 Project Integration Artifacts in Project Preparation Phase

In project integration, two of the most important project artifacts are the project charter and project management plan. The project charter is developed in the Initiation Phase of the project, and its sign-off is one of the most important exit criteria for that phase. The project management plan plays a similar role to the charter document but is drawn up in the Project Preparation Phase. Completion of the project management plan serves as one of the exit criteria for project preparation.

As the project charter was described in the earlier section, the focus here will be on the project management plan. The project management plan serves as the blueprint for effectively managing projects. As a master plan, it includes, through consolidation and integration, a number of component plans. These include (but are not limited to) the scope management plan, schedule management plan, resource management plan, cost management plan, stakeholder management plan, communication management plan, supply chain management plan, risk management plan, quality management plan, conflict resolution plan, governance roles and responsibilities, and operation transition plan. In addition to these subsidiary plans, the project management plan can also include the following components:

- **Change management plan.** Procedures for evaluating, authorizing, and implementing changes to the project
- **Configuration management plan.** To create an auditable trail of important project information
- **Performance metrics baseline.** For identifying and managing key performance indices
- **Management review or project governance.** To establish processes for leadership and guidance with the steering committees and sponsors
- **Development approach.** For communicating the specific approach to project development or execution

The project management plan is predominantly used in the predictive project management. For Agile project management, somewhat irrespective of the methods, there are equivalent considerations. For example, traditional change management is not as applicable in some Agile methods. Instead, Agile project management such as Scrum utilizes a product backlog that is regularly updated with both new ideas and their priorities. Other preparation tasks such as managing risk, agreeing on performance metrics, managing documentations and knowledge, and governance should be important to Agile projects too.

8.3 Managing Project Work during Project Implementation

After developing two of the most important project management artifacts, the project charter and project management plan, the most of the project management effort involves putting them into practice and updating them as required or as situations change. The process of managing project work is where the preparation and planning transforms into action and implementing any necessary changes in order to achieve project objectives. Here, project managers are responsible for executing the project plans, leading project teams, performing project work, managing changes and issues, and effectively communicating and reporting to stakeholders. These responsibilities require integration activities such as communication, analyzing dependencies, managing cross-team issues and risks, knowledge, stakeholders, project metrics, and reporting across the entire project. Given the scope of the responsibility, on larger projects, there are likely to be subteams each working on a particular component of the project. For example, in a large software development project, there will probably be different team leads for business requirements, technical design and engineering, development, testing, deployment, and training. Project managers are also expected to make holistic decisions that optimize the entire project performance and not only specific teams. Often, this requires trade-offs, which are difficult decisions to make, and requires strong project leadership. Change requests are initiated as needed in order to prevent, correct, or repair issues in the execution of the project. This is where the need to manage project knowledge becomes very important.

8.3.1 Managing Project Knowledge

With vast amounts of knowledge being created, modified, and consumed on projects, it becomes important for project managers to manage this information effectively. This process uses organizational knowledge that has already been gained in order to improve the likelihood of project success and to create and share new knowledge gained through the project implementation. This process recognizes the importance of both explicit (tangible) knowledge and tacit (intangible) knowledge.

The process of managing project knowledge takes in the project management preparation artifacts available such as the project management plan, project documents, project deliverables, environmental considerations, and organizational assets and transforms them into specific outcomes – in this case, explicit knowledge managed in a structured manager. The transformation process involves tools and techniques such as sound judgment, data management, communication skills and techniques, and process management. A useful tool within knowledge management is the DIKW pyramid, which explains the transition from *data*, to *information*, to knowledge, and ultimately to *wisdom*.

- **Data** is raw content, usually free from any context. For example, a mechanical engineer service rate of $100/hour is data.
- **Information** is data that has been placed in proper context, often with a comparative analysis. When considering the rate for the mechanical engineer, it is about 10% over the budget of $90/hour. This transforms data into more valuable information.
- When information is further distilled to become more insightful or actionable, it is considered to be **knowledge**. For example, if there is a short supply of Mechanical Engineers in the market, which has pushed up the market-related rate to a premium, then the overbudget rate that has been quoted is now more likely to be approved by the project sponsor.
- Finally, **wisdom** is the assimilation of knowledge and insights into a powerful combination of intelligence. For example, wisdom would be the recognition that the market demand for Mechanical Engineers should be reviewed before agreeing on the baseline budget in future plant construction projects.

8.4 How to Control Performance and Manage Change

During the Implementation Phase of the project, project managers should be regularly monitoring the project performance, creating or assembling performance reports, sharing these with the relevant stakeholders, preparing to facilitate workshops to resolve problems and difficulties, monitoring risks and changes to their statuses, managing change, and intervening as necessary to control project progress and meet the project objectives as agreed to in the project management plan. Once complete, project managers are then responsible for closing the project or phase.

8.4.1 Monitor and Control Project Work

Throughout the entire project but typically with more intensity during the Project Implementation Phase, project managers monitor and analyze project progress, make adjustments to better control the project, and report on progress. By identifying the current project state, determining actions to address performance issues, and forecasting future cost and schedule status, project managers have a much higher chance of completing projects successfully.

Earlier project management artifacts created in the Preparation Phase are highly leveraged at this point. These artifacts include the project charter, project management plan, integrated master schedule, previous progress reports, and other project documents. The project management plan serves as a baseline for work performance. The project charter serves as the guidepost that clarifies the project direction, priorities, and constraints. By leveraging these artifacts, project managers should compare actual project results achieved with the planned performance goals, understand variances and trends, develop options where intervention should be required, take appropriate actions, such as corrective actions to repair defects or preventative actions to mitigate risks, and conduct integrated project change to adjust the course of the project when the original performance goals are no longer sensible or viable. To do this, project managers draw on tools and techniques such as data analysis, sound judgment, decision-making especially on difficult trade-off choices, and meetings to develop action plans.

8.4.2 Perform Integrated Change Control

In research conducted across nearly 450 projects, Dov Dvir and Thomas Lechler found that the positive total effect of the quality of planning is almost completely overridden by the negative effect of goal changes.[1] This implies that it is more important to plan for change in projects, than it is to plan for the project itself. This is why the change control process is considered one of the most vital processes in the Implementation Phase of a project in traditional project management. This is the process of reviewing change requests in an integrated fashion across multiple considerations, such as schedule, scope, and resources; analyzing change requests for impact and cost; developing options to best address the changes; approving (or not approving) changes and communicating the decisions to the relevant stakeholders; managing changes to project deliverables; and updating relevant project documents, such as the project management plan.

Tool. In the Appendix A of this book, the author included an example of a project change request form. See Template 10: Project Change Request Form on page 276.

---

[1] Dvir, D. & Lechler, T., 2004. Plans are nothing, changing plans is everything: The impact of changes on project success. *Research Policy*, 33(1): 1–15.

Projects frequently experience change, especially complex projects operating in turbulent environments. Change requests can be related to the project management process and project deliverables and can vary in the degree of change and control based on project size, complexity, the project organization, and sometimes the external environment (for example, macroeconomic situations, the competitive environment, or industry change). These changes often affect multiple areas of the project, such as the project scope, schedule, project management plan, or any variety of project documents and components. Unless project managers establish a proper change management approach, project success can be negatively impacted by uncontrolled change. For example, an important project team member who falls ill and is absent from work can result in project schedule delays, or additional costs, as others are required to work overtime to catch up the work. Once the project has established baselines, any changes should go through the formal integrated change control process.

8.5 Integrated Transitioning and/or Closing Projects

The final process in integration management is Project Transition and/or Closure. This is the process of finalizing all project activities. As projects are temporary endeavors, all projects must end. An optimal ending or "a happy ending" is when work is completed on the project or in a phase of the project, and the project manager is able to wrap up all the project activities and bring the project to a closure. In this process, the project manager will finalize all the work, update the relevant documents, and conduct lessons learned. Depending on the success of the project deliverables and performance of the project management goals, celebration is often in order.

However, projects can also end for more abrupt reasons. Projects can be terminated early, due to changes in organizational needs. Some projects run out of funding. Other projects, zombie projects, should end, but do not. These are projects that limp along without clear direction or contribution to organizational goals.

At the transition or closure of a project, it is important that project managers evaluate whether the project met and/or exceeded the success criteria as defined in the business case or project charter, whether the deliverables and performance goals were met or exceeded based on the planned goals in the project management plan, and whether all relevant stakeholders are satisfied. Project managers should also assess how well the vendors performed, how effective risk management and change management was, and whether the project has received all the final sign-offs.

Closing the project requires a review of the key project management artifacts created during Initiation and Preparation Phases and updated throughout the Implementation Phase. Depending on the project and the sponsoring organizations, the project team may need to update these artifacts with the latest information. Examples of these artifacts include project charter, project management plan, master project schedule, accepted deliverables, business documents, agreements, procurement documentation, and other project documents.

8.5.1 Postproject Evaluation

Another popular tool for managing project knowledge is the adoption of a PMIS. PMIS is a tool, often driven by technology, that is used to create, store, disseminate, and eventually terminate project information. This can include simple tools, such as a tracking sheet and filing cabinet, or even some combination of emails, shared folders, and Excel sheets. More advanced tools can include technology such as Microsoft® SharePoint®, collaboration tools such as Basecamp, or enterprise grade tools such as Oracle® Primevera and SAP® Advantage Database Server®. However, often more important than the tool itself is a process-centric culture of using the tool, and the business processes and intelligence derived from those tools.

A good practice for project managers is to capture the lessons learned in a register, especially on larger and more complex projects. Lessons learned is sometimes referred to as postproject evaluation, project postmortem, after-action analysis, or postproject evaluation. Regardless of the name, the key goals of a lessons learned register are to understand what went well and should be repeated in future projects, to discuss potential improvements so that future projects can avoid similar challenges, and to strive toward continuous improvement. Some lessons learned also focus on project performance metrics and a comparative analysis of planned versus actual results and activities. Some organizations also conduct information sessions to share project knowledge in a forum, such as a "lunch and learn".

Tool. This book includes a sample template for postproject evaluation in Appendix A. See Template 11: Post Project Evaluation Template on page 278.

8.6 Project Management in Motion

This book is intended for a wide range of audiences and across various industries and functions. Therefore, to balance between competing demands of ease of use, comprehensiveness, modularity, flexibility, and upgradeability, the author has adopted a modular format to this book. This includes the relatively short chapters on key topics. In addition, the appendices include three vital sections:

- Appendix A contains a list of commonly used project management templates for both predictive and adaptive project management approaches.

® Microsoft® SharePoint® are registered trademarks of Microsoft Corporation in the United States and other countries.

® Oracle and Java are registered trademarks of Oracle and/or its affiliates. Other names may be trademarks of their respective owners.

® SAP® Advantage Database Server® are the registered trademarks of SAP SE in Germany and in several other countries.

- Appendix B contains an integrated case study based on a fictitious global company undertaking a number of projects and confronting various project challenges. The case maps closely to the chapters and sections, key concepts, and relevant tools of this book. The goal is to provide instructors, students, and practitioners with a realistic project example to practice and apply the key learnings in the book. For more information and additional case studies in which the author plans to create over time, including potential collaboration opportunities, visit www.optimizepm.com.
- Appendix C contains a glossary of selective terms, reprinting with permission from Dr. Te Wu and Mr. Brian Williamson's book titled "*The Sensible Guide to Key Terminologies in Project Management*", iExperi Press, Montclair, NJ.

Chapter 9

Stakeholder Management – Knowing the People

Summary

Complex and large projects fail at an unacceptably high rate. There are many reasons, but one of them is the improper management of people who are involved in projects. Studies have shown that projects with strong project stakeholder management are more likely to be successful. This chapter discusses four important processes including identifying stakeholders, planning stakeholder engagement, managing stakeholder engagement, and monitoring stakeholder engagement.

Stakeholder is a relatively new performance domain for project management even though the management study of stakeholder became popular with Freeman's (1984) book titled *Strategic Management: A Stakeholder Approach.* This was a pioneering work on the subject of stakeholders, and when it became widely accepted in management, an emerging field called stakeholder theory was formed. In project management, stakeholder was traditionally discussed in conjunction with communication. But clearly, stakeholder is much more than communication. Today, as projects have grown in complexity, stakeholder management provides additional lens through which project professionals can understand project scope, analyze project environment, and manage and shape stakeholder expectations.

This chapter addresses these three important questions pertaining to stakeholder management:

1. Why is managing stakeholders important?
2. How to identify stakeholders?
3. How to best engage stakeholders and manage their expectations?

9.1 Importance of Engaging Project Stakeholders

Projects involve people: those who work on the project, those who benefit from the project, and those who in some way impact or are impacted by the project. These people are called stakeholders, and each stakeholder has certain roles and expectations about the management of the project and the achievement of project deliverables. The larger the number of project stakeholders, the more emotions, personalities, perspectives, and complexities are added to the challenging task of coordinating multiple expectations and roles. The potential for conflict becomes greater. However, stakeholders also bring certain advantages, such as ideas, problem-solving capabilities, and more experience. The project manager is tasked with managing stakeholder roles and expectations such that the advantages are leveraged, and the risks of conflict are mitigated.

Stakeholders can be internal or external to an organization and have varied levels of involvement over the duration of a project life cycle. Internal stakeholders often include sponsors, project and functional managers, project team members, subject matter experts, fund providers, executives, shareholders, and/or general company employees. External stakeholders can include suppliers and business partners, government institutions, creditors, customers, professional bodies, unions, the media, competitors, and the broader society. Knowing who the project stakeholders are, and managing them accordingly, is critically important to the successful planning and execution of projects. Research has found that the influence of both primary and secondary stakeholder engagement has a significant impact on project success.[1] Furthermore, different stakeholders can have a different impact on the success of specific project life cycle phases.[2] On large and complex projects, process success spans beyond the traditional approach of meeting allocated time, cost, and scope.

---

[1] Beringer, C., Jonas, D., Kock, A., 2013. Behavior of internal stakeholders in project portfolio management and its impact on success. *International Journal of Project Management*, 31(6): 830–846.

[2] Aaltonen, K., Kujala, J., 2010. A project life cycle perspective on stakeholder influence strategies in global projects. *Scandinavian Journal of Management*, 26(4): 381–397.

Success is also indicated through more subjective parameters, such as customer satisfaction, business success, and meeting stakeholder expectations. However, not all project stakeholders have similar interests in a project. Some stakeholders can be supportive or neutral, whereas others can distract and obstruct project execution when they are not invested in its implementation. For example:

- Organization executives are usually interested in the project benefits, but not necessarily the details of execution.
- Project sponsors are interested in the outcome, but not always the risks and obstacles.
- Functional managers are predominantly interested in the use of their functional resources and decisions affecting their functional areas.
- Project managers want to meet time, budget, and scope constraints.
- Customers and end users care about the outcome and output and the associated benefits.
- Society may have no direct interest, but indirectly, our society often benefits from innovations that come from competition.

It is important to understand different stakeholder interests and the ways in which they can impact the project, because without stakeholder support and buy-in, project process success and project deliverable success are threatened.

Stakeholder management is the processes of identifying, evaluating, and engaging the people, groups, or organizations that may affect the work of the project and may be impacted (real or perceived) by the project output. The four major activities include the following:

1. Identifying stakeholders (Ideation and Initiation Phases)
2. Planning stakeholder engagement (Preparation Phase)
3. Engaging stakeholder (Implementation Phase)
4. Transitioning and/or closing stakeholder engagement (Transition and/or Closure Phase)

As the project leader, project managers often have a primary responsibility conducting these four activities. In identifying stakeholders, project managers will work with project sponsors and the core project team members to identify people and parties who have stakes in the project. Project managers will evaluate and plan stakeholder engagement by analyzing the stakeholder community and developing an appropriate plan of action to secure and maintain their support. Finally, project managers are relied upon to manage and monitor stakeholders by building relationships of trust and commitment and regularly monitoring stakeholder changes, support, and contributions. By managing stakeholders carefully throughout the project, the project benefits from a higher probability of success.

9.2 Identifying Stakeholders

There is no project without project stakeholders. However, it can be challenging to identify all project stakeholders upfront. Furthermore, project stakeholders and their related interests can change throughout the life cycle of the project. For this reason, identifying stakeholders begins early – as early as Ideation Phase. By the Initiation Phase, especially during the development of the project charter, stakeholder identification becomes a crucial process. Identifying stakeholder can occur throughout the project life cycle as needs arise. These needs include stakeholders working on various phases of the project, change in governance or leadership, customer preference changes, or when the external environment shifts such as a competitor introducing a new product.

During Ideation and Initiation, identifying stakeholder can help with the development of ideas into business ideas. Other artifacts that help with determining stakeholder include business agreements, environmental factors, and other planning documents. Working with these artifacts, project managers apply their business judgment, gather and analyze data, conduct meetings and interviews, and apply other techniques to identify stakeholders. Project managers can also create a stakeholder register to start capturing the relevant information and new discoveries. This process of identifying stakeholders should be repeated at the start of each project phase or with every significant project or organization change.

9.2.1 Stakeholder Management Tools

There are a number of tools and techniques available for collecting, analyzing, and interpreting information for identifying stakeholders. The stakeholder evaluation matrix and power/interest matrix are two particularly simple but effective tools.

Tool: Stakeholder Evaluation Matrix (Figure 9.1). This figure shows the Stakeholder Evaluation Matrix from PMO Advisory. In analyzing the information from inputs and identifying stakeholders, the project manager must consider whether a stakeholder has any expectation of the project and what that expectation might be. The project manager should also consider whether the stakeholder would experience any real or perceived impact as a result of the changes introduced by the project, together with the extent of emotional impact. More importantly, if nothing is done to manage the stakeholder, the extent of negative impact that this could have on the project should be considered, i.e., how much influence does this

Internal or External Stakeholder	Project Role	Evaluation Questions (Use a scale of 0 to 5, where 0 is non-existent, 1 is low, and 5 is high)					Score = Sum (Q1:Q4)*Q5
		Q1: Does this stakeholder have any expectation of this project?	Q2: What is the magnitude of real or perceived impact of change to this stakeholder?	Q3: What is the extent of "emotional" impact to this stakeholder?	Q4: If nothing is done to manage this stakeholder, what is the extent of negative impact to this stakeholder?	Q5: How much influence does this stakeholder have on your project?	
Internal							
Internal							
Internal							
Internal							
Internal							
Internal							
Internal							
Internal							
Internal							
External							
External							
External							
External							
External							

Figure 9.1 Stakeholder evaluation matrix.

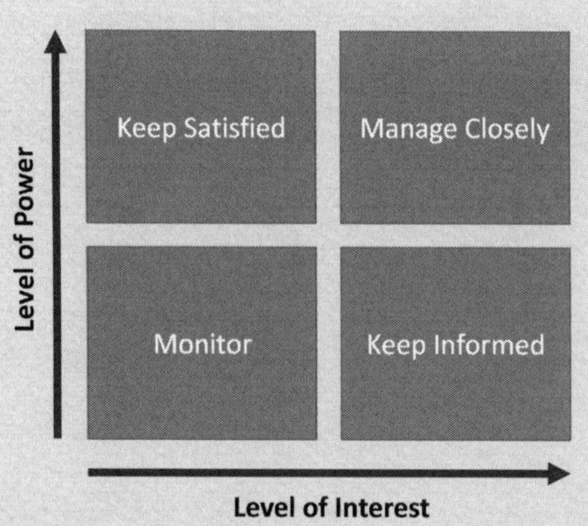

Figure 9.2 Power/interest matrix.

stakeholder have on the project? There are a number of ways of calculating the final score. At PMO Advisory, their score is based on this formula: (Q1+Q2+Q3+Q4)*Q5, with each question scored on a scale of 0–5, where zero is nonexistent, 1 is very low, 2 is low, 3 is moderate, 4 is high, and 5 is very high. The highest score is 100.

Tool: Power/Interest Matrix (Figure 9.2). This 2×2 matrix assists project managers in categorizing project stakeholders across two dimensions, namely, level of power and level of interest. The four resulting categories are as follows:

1. **Manage closely.** These are important stakeholders that project managers must monitor carefully. Often, special stakeholder engagement activities are planned for this group of stakeholders.
2. **Keep satisfied.** These stakeholders are not particularly interested in the project, but it is important to keep them satisfied. They can be powerful allies for decision-making and resource allocation. Where possible, project managers should leverage existing stakeholder activities to manage this group. But if there is a high number of these stakeholders or if there are other important factors, then project managers may design special engagement activities.
3. **Keep informed.** These stakeholders are highly interested in the project, even though they may not have much decision-making power. They should be included in the communication plan. Project managers should also use existing engagement activities to keep them informed.

4. **Monitor.** These stakeholders are on the peripheral. Project managers should include them in routine communications and monitor their support and commitment if there are any changes.

9.3 Planning Stakeholder Engagement

After identifying all relevant stakeholders, the next process is the planning of stakeholder engagement. This process draws on the preceding process of outlining the approach of interaction with stakeholders based on their needs, interests, expectations, and influence[3] and is performed periodically throughout the project life cycle. As with the previous process, planning for stakeholder engagement takes place early on in the project life cycle but should undergo multiple iterations to reflect the dynamic nature of stakeholder engagement. As the project progresses, the stakeholder community will also change. Stakeholders may transition in and out of the project, or their information needs may change during the project. The inclusion of new stakeholders, the exit of current stakeholders, changing stakeholder needs, organizational structural changes, the start of a new project phase, or changes to other project processes may all create a need to revise the stakeholder engagement plan.

Planning for engagement can be an intense project process, as it may require the collective inputs of the project team during the Preparation Phase. The project manager should gather the previously created project management artifacts such as the stakeholder register, business case, project charter, project management plan components, and project documents and agreements. By analyzing the data collected, working with the project team members on how to best engage the stakeholders, project managers can create the stakeholder engagement plan.

Tool: Stakeholder Engagement Plan. See Figure 9.3 for an example from PMO Advisory. This document provides strategic guidance for productive stakeholder engagement during project execution and monitoring and forms a component of the project management plan.

---

[3] Project Management Institute. 2017. *A Guide to the Project Management Body of Knowledge (PMBOK® Guide) (6th Ed.)*, Project Management Institute, Newtown Square, PA.

Stakeholder Engagement Action Plan

Initiative Name:	
Evaluation Date:	

Evaluator's Name:	
Evaluator's Role:	

#	Stakeholder / Group	Score	Role (from the project perspective)	Status - Blocker, Pessimist / Disinterested, Supporter	Level of Support Required (Necessary, Desirable, Unnecessary)	Stakeholder Needs / Expectations (What do they want from the Project Team?)	Project Needs / Expectations (What the Project needs from the stakeholders? What level of commitment / support is necessary?)	Engagement Required? (Timing / Frequency)	Responsible Parties for Engagement
1									
2									
3									
4									
5									
6									
7									
8									
9									
10									

Figure 9.3 Stakeholder engagement plan.

9.4 Engaging Stakeholder

The effective engagement of stakeholders can improve communication and reduce project risk and associated costs, thereby increasing the likelihood of project success. Engagement means sharing project goals, benefits, and risks, as well as the value of stakeholder contributions. When stakeholder satisfaction is accepted as a project objective, its achievement becomes particularly reliant on sound management practices that are tailored to the stakeholder community. This is a continuous process that involves communication and collaboration with stakeholders in order to gain support and minimize resistance. Consulting those stakeholders who will be affected by the project outcomes or deliverables in a process of cocreation can improve stakeholder investment and acceptance of the project. Interacting with stakeholders frequently at the appropriate phases of the project life cycle, and creating a sense of inclusion, often improves commitment to the project.

Project managers should engage key stakeholders to develop trust through sharing the project vision, new learning, enjoyment, failures, and success; aligning interests and motives; recognizing differences and arriving at compromises; encouraging bidirectional communication and listening; using appropriate decision-making strategies; focusing on tasks and issues, rather than personalities and people; strategically selecting which battles to fight; and managing stakeholders using the plan created in the stakeholder engagement planning process. Project managers are responsible for a shared and sustained understanding of project goals, objectives, benefits, risks, and contributions among the stakeholder community. This is primarily achieved through communication.

Project managers are required to communicate and negotiate around stakeholder responsibilities and expectations in order to manage conflict and disagreements, process issues, review performance issues, and other concerns. Furthermore, the project manager should be responsive and proactive in anticipating future issues. By leveraging the stakeholder management artifacts such as the stakeholder registry and stakeholder engagement plan as well as other project management artifacts such as the project charter, project communication plan, and project management plan, project managers can proactively engage the key stakeholders and nurture their support for the project.

9.4.1 Monitor Stakeholder Satisfaction

One important aspect of engaging stakeholder is the regular monitoring of stakeholder satisfaction. By noting changes in stakeholder satisfaction, project managers can adjust strategies for engaging stakeholders throughout the project life cycle in order to acquire, confirm, or retain their commitment. This is a continuous process that spans the project life cycle in order to review and improve the effectiveness and efficiency of stakeholder engagement on the project. Regular monitoring of stakeholders is important in order to understand the impact of changes to the project environment, project scope, or organizational structure, on project stakeholders. Similarly, changes to stakeholder interests, the introduction of new stakeholders, and stakeholder attitudes toward the project should be monitored for their impact on the project process and deliverables. During this process, the project manager considers the value of engagement with each stakeholder or stakeholder category, including the positive value (benefits derived) of stakeholder support and the negative value (costs) associated with the failure to engage certain stakeholders effectively.

9.5 Transitioning and/or Closing Stakeholder Engagement

As the project comes to a close, it is important for project managers to thank all stakeholders graciously for their support and contribution to the project. Even if selective stakeholders are less helpful, this is a good time to mend fences and build bridges in preparation for the next project. Key stakeholder should be invited to attend team-wide project closure or transition activities, such as the project celebration. On projects that have activities after closure, it is common for some stakeholders to continue to play integral roles in managing operational activities or in the next phase of project implementation. Most importantly, as a part of lessons learned, project managers should highlight the stakeholder management activities and techniques that contributed to the project success so they can be repeated in the future.

9.6 Project Management in Motion

This book is intended for a wide range of audiences and across various industries and functions. Therefore, to balance between competing demands of ease of use, comprehensiveness, modularity, flexibility, and upgradeability, the author has adopted a modular format to this book. This includes the relatively short chapters on key topics. In addition, the appendices include three vital sections:

- Appendix A contains a list of commonly used project management templates for both predictive and adaptive project management approaches.
- Appendix B contains an integrated case study based on a fictitious global company undertaking a number of projects and confronting various project challenges. The case maps closely to the chapters and sections, key concepts, and relevant tools of this book. The goal is to provide instructors, students,

and practitioners with a realistic project example to practice and apply the key learnings in the book. For more information and additional case studies in which the author plans to create over time, including potential collaboration opportunities, visit www.optimizepm.com.

- Appendix C contains a glossary of selective terms, reprinting with permission from Dr. Te Wu and Mr. Brian Williamson's book titled "*The Sensible Guide to Key Terminologies in Project Management*", iExperi Press, Montclair, NJ.

Chapter 10

Scope Management – Defining Scope and Determining Requirements

Summary

This chapter focuses on the "what" of the project, more formally known as the project scope. Scope is central to any project, as fulfilling the scope is the primary reason for the existence of the project in the first place. While seemingly straightforward, managing scope well, especially in a changing environment, can be one of the greatest project challenges. So much so that the recent advancement in the Agile project approach is to tackle the very issue of defining and managing scope. The importance of carefully defining project scope, identifying the specific requirements, creating a work breakdown structure (WBS) to guide and direct work activities, and validating and controlling scope throughout the project will be discussed in turn.

In traditional project management, especially before the recent advancements in stakeholder management, project professionals often initiate projects by addressing the question of scope. Logically, this makes intuitive sense – understanding "what" is foundational to the planning of cost, resources, communication, and most other project management domains. However, given the challenges of defining scope, project professionals may utilize different methodologies to identifying, analyzing, and managing scope based on the project deliverables, project environment, and other important factors. Here, some projects may naturally gravitate toward the more adaptive and Agile approaches while favoring the predictive and traditional methods.

This chapter addresses three important questions with regard to scope management:

1. Why is scope management important?
2. What are the challenges of managing scope?
3. How to manage scope well on projects?

10.1 What Is Scope Management?

The scope of a project is the sum of all work required to deliver and manage a project successfully. Broadly, there are two types of scope: **(1) project scope** refers to project deliverables, such as features and functions of a product, and can also include the desired results, such as increasing the market size by 10%; **(2) project *management* scope** includes the work required to deliver a product, service, or results, including all the project management activity efforts. Total scope includes both the project scope and project *management* scope.

Scope management comprises multiple processes and activities for ensuring that a project includes all the work required, and only the work required, to complete the project successfully and with effective management. Project scope management includes four major activities, namely:

1. Conceiving initial scope (Ideation and Initiation Phases)
2. Planning, defining, and validating scope (Preparation Phase)
3. Achieving and controlling scope (Implementation Phase)
4. Transitioning and/or closing scope (Transition and/or Closure Phase)

The deliberate management of scope is vital to project success, but scope is not always easy to determine. The clarity of scope can be influenced by a number of factors, including systems, processes, requirements, organizational challenges, communication, and stakeholders.[1] See page 60 for a discussion of fuzzy front end, an acute and particular problem that can occur on nearly all complex projects. One of the primary assumptions of scope management is that project stakeholders, such as the sponsor or customer, are able to identify and articulate the basic requirements of the project in advance. The project team is then responsible for collecting these requirements and, with further analysis, is able to create detailed requirements and

[1] Kumari, N., Pillai, A. S., 2014. A study on project scope as a requirements elicitation issue, in *Computing for Sustainable Global Development (INDIACom)*, IEEE 2014 International Conference on 2014 March 5–7, New Delhi, India, pp. 510–514.

specifications. However, it is often challenging, and sometimes impossible, to identify, define, and plan for all requirements in advance. This is particularly true when the competitive environment changes so frequently and so quickly that plans are likely to become obsolete, regardless of how well the scope has been defined. The difficulty for project managers is to manage these unknowns, and the potential for scope changes effectively.

EXAMPLE: THE CHALLENGES OF SCOPE MANAGEMENT

Example 1: A university is looking to construct a new building in order to meet the growing need of a larger student body. After working with key stakeholders, such as school administrators, professors, and students, it is determined that the building will be 150,000 square feet, 5 floors with a large ground floor, and 45 classrooms. These are the product requirements. However, after breaking ground, it is discovered that the condition of the ground soil cannot support such a large building. Thus, the requirements cannot be met.

Example 2: A school is undergoing technological updates and adopting new software. They would like to prepare the technology in the building for the next 10 years. However, technology improves at an exponential rate, which makes a 10-year plan rather difficult. The relevance of technology for the next year may be reasonably forecasted, but the technology available in 10 years' time may not even be comprehensible. If the project manager does attempt to plan for a 10-year technology update, then it is very likely that the customers (school's faculty, staff, and students) will continuously ask for new features, and the scope will slowly and gradually shift or expand (scope creep if not managed well). Thus, the requirements or scope may continuously change.

10.1.1 Managing Scope in a Turbulent Environment

The environments in which organizations operate today can be turbulent, dynamic, and constantly changing. This makes it especially challenging to plan for project scope. To meet these challenges, project managers must identify and implement the project management approach most suitable for the project challenges (e.g., waterfall versus Agile versus hybrid). To avoid scope creep, project managers must follow a consistent change process, including steps to identify, analyze, categorize, document, approve, prioritize, and communicate results. Scope at the detailed level is often the most difficult to define and plan for, so it becomes crucial to utilize the tools and skills necessary to develop a detailed analysis. Project managers should drive as much clarification as possible, both early on during planning and throughout the project life cycle, as necessary. For project managers, this is where the art and science merge within the project management discipline.

10.2 Conceiving Initial Scope

The beginning of any project starts with an idea. Depending on organizations, idea generation can be spontaneous and serendipitous, formal with a disciplined process of conceiving and developing ideas, or somewhere in between. The ideas themselves can be in response to addressing existing gap or resolving problems or completely innovative to stretch the existing boundary. In the very first step of forming the idea, the idea itself serves as the initial scope.

As the ideas are captured, developed, nurtured, and transformed into business cases, the scope becomes more concrete and takes definitive shape. Along this path, boundaries start to appear, and real-world considerations such as constraints of time and budget start to influence the ideas itself. Near the end of the Ideation Phase, project scope is usually sufficiently well defined that business decisions can be made. At this phase, however, the scope is still abstract, and much more work would be required before implementation starts.

Upon approval of the business case, the project moves to the Initiation Phase. In this phase, the scope, including the project scope and the project *management* scope, starts to solidify. As project managers acquire the core team and work on the project charter, the project team is likely to examine the high-level scope closely and start to develop the next level of details, often called the business requirements.

Definition. A requirement is a condition or capability that is required to be present in a product, service, or result to satisfy a contract or other formally imposed requirement.[2]

Working with the project team, project managers or business analysts start to create a business requirement document, which is a document that identifies and lists all the business level needs and wants of a project. For example, in the software development life cycle, business requirements are further refined, analyzed, and transformed into functional and nonfunctional requirements in the Preparation Phase.

10.3 Planning and Defining Scope

An unclear scope can lead to misunderstood expectations, dissatisfied stakeholders, and wasted time, as the project team works to achieve the wrong deliverables, in the wrong way. To ensure the proper management of scope, project managers must first develop and define an overall approach, including how to gather details

[2] Williamson, B. & Wu, T., 2019. *The Sensible Guide to Key Terminologies in Project Management*, iExperi Press, Montclair, NJ. Glossary.

and requirements, document and validate the requirements, obtain sign-offs, and manage and maintain the requirements, especially when there are changes. In an Agile approach, the project manager must also consider how to enhance project scope requirements continuously, as new situations emerge with new ideas. In predictive or traditional project management, the plan and defining scope process generally takes place once during the Preparation Phase. In Agile approaches, planning and defining scope occurs repeatedly at various points in the sprints or iterations.

To plan and define scope, project professionals, often the business analysts, start with the scope and requirement-related artifacts from the previous phase such as the business case, project charter, and project management plan to analyze and dive deeper into "how" the project scope will be implemented. The project manager, working with the subject matter experts and the business analysts, also defines the project management work required to deliver the project successfully. This requires a combination of expert judgment, data analysis, and meetings with key stakeholders to define both the project and project management scope. For larger projects, the project managers are encouraged to create the project scope management plan, which documents how the project and project management scope will be defined, monitored, controlled, and validated, and the requirements management plan, which describes how project and product requirements will be analyzed, documented, and managed.

In some publications, such as the *PMBOK® Guide*, the scope management plan is a component of the project management plan, a comprehensive plan that details how project managers will manage their projects. The scope management plan can be a high leveled or detailed, formal or informal, document, depending on the needs and complexity of the project. Components of this document include a process for developing the project scope statement (part of the scope baseline), a process for the creation of the WBS and a WBS dictionary (part of the scope baseline), procedures for approval and maintenance of the scope baseline including change control processes, and procedures for the formal acceptance of project deliverables. The requirements management plan, also known as the business analysis plan, also forms part of the project management plan. The requirements included in this document can be planning, tracking and reporting activities, configuration activities, such as change management, processes for prioritization of requirements, metrics and the rationale for their use, and structure and guidelines for the requirements traceability matrix. This document is vital in the closing project phase, as customers use this document as a reference to what was agreed upon and what was actually delivered.

10.3.1 Gathering Requirements

Upon completing the scope management plan, the project manager and/or the business analyst are prepared to gather requirements, which is a process of identifying, analyzing, documenting, and capturing stakeholder needs and requirements to meet project and project *management* objectives. The purpose is to achieve a mutual understanding of the project and project *management* objectives among the project team.

Requirement documents can take many forms, such as high-leveled or detailed business requirements, stakeholder needs, requirements about solution features or functions, product behavior and functional requirements, nonfunctional requirements, such as the qualities required for the product to work effectively, and other requirements such as transition, organizational readiness, and quality, and project process requirements. In a traditional waterfall approach, this process is typically completed once at the beginning of the project. Additional requirements are subsequently managed as changes or placed on hold until the next project phase. In an Agile methodology, the process is often more iterative, taking place repeatedly throughout the project life cycle.

To aid in the collection and analysis of requirements, user stories are becoming more popular especially in the adaptive or Agile methods. A user story is an informal, natural language description of one or more features of a software system. User stories are designed to provide clarity and real-life use of how the user will interact with the system.[3] By providing this contextually rich background, requirements are more relatable. For complex projects, such as application development, this approach to gathering requirements has proven to be invaluable.

Tool: User Story. See Template 4: User Story on page 272 for a sample template.

10.3.2 Defining and Refining Scope

After gathering the project scope, the next step is to define (and later to refine) scope. This is the process of creating a detailed narrative of the project and project management activities. The key benefit of this process is that it describes the product, service, or result boundaries and acceptance criteria. The define and refining scope process draws on information from the requirements documentation and scope management plan, among other inputs such as the project charter, project management plan, and other project planning documents.

Project managers apply a range of tools and techniques and work with business analysts to finalize the scope definition. These tools include data analysis (to ensure that requirements are realistic), decision-making (especially important in more contentious projects where serious trade-off decisions are required), interpersonal and team skills (solidifying scope often requires strong negotiation and

[3] Williamson, B. & Wu, T., 2019. *The Sensible Guide to Key Terminologies in Project Management*, iExperi Press, Montclair, NJ. Glossary.

teamwork skills), and product analysis (detailed analysis to ensure that the product will meet customer or market demands).

For a traditional project management approach, defining scope is a linear process with incremental steps, including the following:

1. Identify and list project deliverables
2. Determine project acceptance criteria
3. Determine which deliverables are within or without the scope
4. Develop scope definition

In an Agile approach, defining scope occurs in iterations throughout the project life cycle, beginning in the Preparation Phase. The user requests specific features and functions; the product manager or Scrum Master then adds these features and functions to the product backlog. The product backlog is reviewed during each Implementation Phase sprint (Agile approach), and the high priority features are selected for implementation. The project team then works together to define and implement the features in the next iteration.

10.3.3 Creating the Work Breakdown Structure

The WBS is a technique for dividing larger project deliverables into smaller and more manageable pieces. This process provides a guiding framework of the work necessary for completing and achieving key project and product deliverables. Although this process is often performed at a single point in the Preparation Phase, creating the WBS process can occur at multiple points throughout the project life cycle. In this case, project managers continue to develop more work details until such points that they are comfortable managing the work with an acceptable level of precision.

Insights from the Front Line. The level of WBS detail often differs between experienced and junior project managers. Experienced project managers are naturally more comfortable at a higher level of detail, whereas the junior project managers require more precise instruction to achieve the same level of confidence.

10.3.3.1 Work Breakdown Structure Components

To create the deliverables, an important technique is to decompose the required project activities methodically into smaller, more specific, and more manageable units. Through this hierarchical breakdown of complex work into controllable units, project managers and project teams can estimate duration, resources, effort, and costs and evaluate risks associated with the more granular unit of work.

This structure serves as a framework for organizing the total project scope into components, some of which are relevant to other knowledge domains. The *work* in the WBS refers to the project and project management deliverables that result from an activity, rather than the activity itself. The lowest level of the WBS for which cost and duration can be managed is the *work package*, where work is estimated, monitored, and controlled. At this level, activities include estimations of effort and duration, schedules, resource assignments, and project controls.

10.3.3.2 Create Work Breakdown Structure Process

Creating a detailed WBS is considered to be one of the most vital tasks in project management, because it provides greater clarity on the definition of work required to implement projects successfully; enhances the visibility of work required, and the credibility of the project planning; serves as the basis for planning in all subsequent project management work, including schedule, resources, cost, quality, and risk; sometimes highlights problem areas; and functions as an important communication document throughout the project life cycle.

Project managers should use this opportunity to create the WBS as a team collaboration exercise to start developing the team. By gathering the core team members representing the various subteams and with project managers, a chief facilitator and mediator, the project team can learn to work with each other. At this early phase of the project, it is common to have disagreements on how to perform certain tasks or how to manage the integration of multiple work streams. Project manager can serve as a leader, facilitator, and mediator during sometimes difficult situations on how to best organize the project work.

After gathering the high-level scope, project managers and business analysts apply their expert judgment and using decomposition techniques to create the structure and its subdivision of components. The *decomposition technique* broadly involves identifying and analyzing the major components, deliverables, and work and dividing these into smaller components (work packages) until all components are the right size. The "right size" will depend on the project and the project manager's expertise. This is the point at which the project manager feels comfortable for conducting a detailed analysis of schedule, resources, effort, and risk and for controlling the project activities. The WBS should be either created in collaboration with the project team or reviewed with and agreed upon by the project team.

WBS can be approached and organized in various ways. A few of the most common approaches include the top-down approach, WBS templates, and bottom-up approach. WBS can also be presented in a number of different formats, such as the organization chart, tree, or mind map. Finally, WBS components and levels can be arranged by main deliverables (Figure 10.1), functional (Figure 10.2), or project life cycle (Figure 10.3).

After completing the WBS and the other planning work products, the scope should be validated before moving to the Implementation Phase of the project.

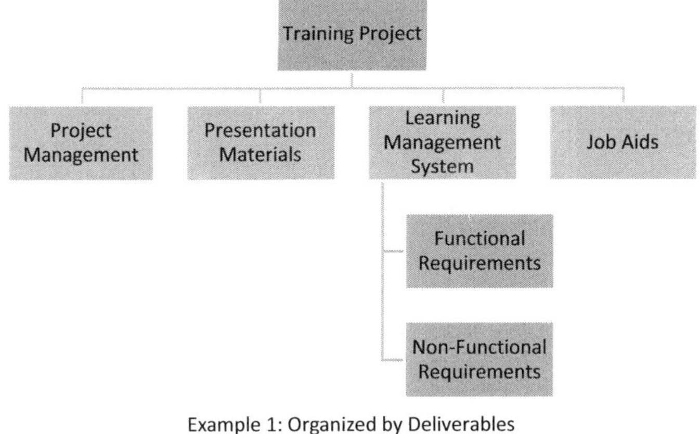

Figure 10.1 WBS by deliverables. WBS, work breakdown structure.

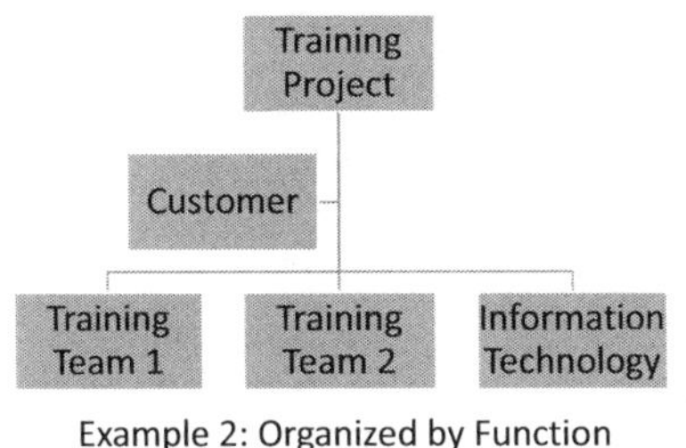

Figure 10.2 WBS by functions. WBS, work breakdown structure.

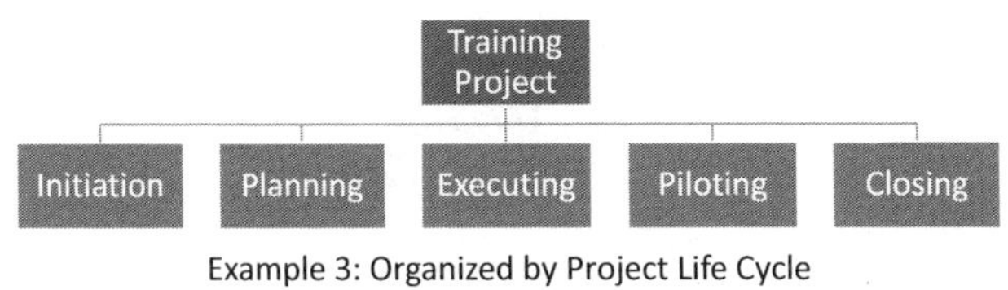

Figure 10.3 WBS by life cycle. WBS, work breakdown structure.

10.3.3.3 Validating Project Scope

Validating project scope process is critically important for increasing the likelihood that project deliverables will be accepted. This process takes place repeatedly throughout the project life cycle and requires careful inspection of the scope planning documents, such as requirements documents, the scope management plan, and the traceability matrix.

In an Agile environment, validating project scope occurs within the project team. The Scrum Master or other project leaders work together with the project team to decide on the requirements from the product backlog.

This formal, objective step to validate project scope provides greater assurance that all stakeholders are in agreement and that the customer or sponsor is likely to accept the project deliverables. By using techniques such as inspection and decision-making on trade-off considerations, the project manager and business analyst reviews the project and the project management scope with the sponsor and the customers to reach agreement and formally sign off on the validated scope.

Note, in Section 16.4 on Managing and Controlling Quality on page 206, there is a discussion of Control Quality. There can be some confusion between these two processes. The primary difference between the control quality and validate scope processes is that the control quality process includes quality checks to ensure that the project deliverable is produced in the *right way*, whereas the validate process checks that the *right deliverable* is produced and accepted by the customer or sponsor.

10.4 Achieving and Control Scope

In the Implementation Phase of the project, project managers must carefully monitor the project and project *management* scope and their changes. Throughout the Implementation Phase, the project manager with business analysts' assistance works closely on the progress of building the project scope and also managing the project as agreed. If and when changes occur, the project manager should work with the impacted stakeholders to evaluate the change and its acceptance or rejection. The project manager then informs the relevant stakeholders of the resulting decision. Depending on the extent of the change, project managers should follow the predefined change management and escalation process to manage the change properly. Through this process, the scope baseline is maintained, and scope creep (the expansion of the project or product deliverables without adjusting timelines, costs, and resource allocations) is avoided.

For Agile projects, instead of managing change using a traditional change management approach, the project team may wish to create a product backlog. A product backlog is commonly used in Agile, and it contains a listing of all activities, tasks, new features, changes to existing features, or bug fixes, which need to be completed to fulfill project or product requirements satisfactorily or to achieve a specific outcome.[4] In addition to serve as a valuable tool that captures ideas, a product backlog can also be a political tool as no idea is truly rejected. It may just be delayed to the next iteration.

[4] Williamson, B. & Wu, T., 2019. *The Sensible Guide to Key Terminologies in Project Management*, iExperi Press, Montclair, NJ. Glossary.

Tool: Product Backlog. See Template 5: Product Backlog on page 272 in Appendix A for an example of a product backlog template.

10.5 Transitioning and/or Closing Scope

Upon completing the project implementation, the focus shifts to closing the scope management process. This is where the project business analyst should critically evaluate the project scope, by comparing with what was agreed and what was delivered. The project manager should focus on the project management scope. Ideally, the business analyst and project manager developed a traceability matrix earlier in the project, and it can now be used to ensure that what was agreed is actually built. Before transitioning the project to another phase or operations or closure, project managers should obtain proper sign-offs and approvals from customers and governance bodies, support teams, and other key stakeholders before formalizing closure.

10.6 Project Management in Motion

This book is intended for a wide range of audiences and across various industries and functions. Therefore, to balance between competing demands of ease of use, comprehensiveness, modularity, flexibility, and upgradeability, the author has adopted a modular format to this book. This includes the relatively short chapters on key topics. In addition, the appendices include three vital sections:

- Appendix A contains a list of commonly used project management templates for both predictive and adaptive project management approaches.
- Appendix B contains an integrated case study based on a fictitious global company undertaking a number of projects and confronting various project challenges. The case maps closely to the chapters and sections, key concepts, and relevant tools of this book. The goal is to provide instructors, students, and practitioners with a realistic project example to practice and apply the key learnings in the book. For more information and additional case studies in which the author plans to create over time, including potential collaboration opportunities, visit www.optimizepm.com.
- Appendix C contains a glossary of selective terms, reprinting with permission from Dr. Te Wu and Mr. Brian Williamson's book titled "*The Sensible Guide to Key Terminologies in Project Management*", iExperi Press, Montclair, NJ.

Chapter 11

Schedule Management – Understanding "When"

Summary

In project management, perhaps there is no stronger association with the project management discipline than with time. Outdated, but the importance is still valid; schedule and time continue to be considered one of the most important areas in project management. Time as a measurement is as objective as measurement can be; time is unidirectional, and it is absolute.

This chapter focuses on the "when" of projects by discussing different ways to plan and create a project schedule. Processes covered include defining and sequencing activities, estimating activity durations and the critical path, developing schedules, and controlling and managing project schedules during project execution. The project management information systems tool, Microsoft Project, is also discussed briefly.

This chapter concentrates on three central questions pertaining to schedule management:

1. Why is schedule management important to project management projects?
2. How to plan and manage time effectively?
3. What to do when there is not enough time to complete the project?

11.1 Importance of Schedule Management

There are 24 hours in a day, 7 days in a week, and 12 months in a year. Time is finite, measurable, and unidirectional, which makes the schedule arguably the most unambiguous component of a project. Time is more definitive and absolute than other project attributes. As time has only one direction, once passed, there is no return. Schedule addresses concerns of when the project will be completed, the latest time at which an activity can start without impacting the overall schedule, the dependencies among multiple activities, consequences of delays along the critical path, stakeholder time commitments, schedule and resource bottlenecks, and the least expensive way to expedite a project. For these and other reasons, project management has traditionally been more heavily associated with schedule and time management than other knowledge domains.

There are four processes in the project schedule management knowledge domain that assists in the timely completion of projects. They are the following:

1. Conceiving initial schedule (Ideation and Initiation Phases)
2. Planning, sequencing, estimating, and validating schedule (Preparation Phase)
3. Controlling and refining schedule (Implementation Phase)
4. Transitioning and/or closing schedule (Transition and/or Closure Phase)

In smaller projects, these processes can become so closely related that they form one continuous process outlined by one or two project managers. Yet, as projects become larger, complexities and interdependencies create the need for scheduling tools to assist in planning, monitoring, and reporting on project progress. Factors such as resources, approach (predictive or adaptive), technology, and project life cycle length and complexity can influence the way in which project managers apply scheduling processes. Where feasible, project schedules should contain some elements of flexibility and adaptability to account for new information and even lessons learned.

11.2 Conceiving Initial Schedule

Conceiving an initial schedule occurs early in the Ideation Phase. Often, one of the first questions pertaining to ideas considered for implementation after defining the initial scope is "how long does it take to complete this initiative?" At the beginning, the notion of schedule is vague as the scope of the initiative has yet to solidify. But throughout the Ideation Phase as the ideas are captured, developed, nurtured, and transformed into business initiatives, the scope and schedule become more concrete and take definitive shape. See Table 13.2 on page 168 for a summary of estimating methods.

Along this journey, customers and sponsors may start to exert their influence, and schedule becomes firmer. Depending on projects, the focus may shift to a rigid completion date, as often the case in regulatory, compliance, or seasonal projects. For example, in universities, most major system changes that impact customers should ideally launch before the start of the school year. Other projects, the emphasis may be on the target start date, to demonstrate to investors of positive momentum toward stated goals. Regardless, schedule during the Ideation Phase often remains elusive, especially at the detail levels.

Upon approval of the business case, the project moves to the Initiation Phase. In this phase, the scope is firmer, and the corresponding schedule can also be more concrete. As project managers acquire the core team and work on the project charter, the project team should start to develop the major milestones with planned start and completion dates. There will probably continue to be significant vagueness at the detailed activity and task level, but the overall project plan starts to take shape.

11.2.1 Planning Schedule Management

For larger and complex projects, project teams may need to "plan for the plan", and planning schedule management is the process of establishing the approach and policies for planning, developing, managing, and controlling the project schedule. Depending on whether a predictive or adaptive approach has been chosen, planning schedule management will take place once at the beginning of the project or iteratively at key points throughout the project life cycle.

Typically, the project manager working closely with the core team examines and evaluates the project management artifacts available, such as the project management plan, project charter, and other project documents. Applying a combination of expert judgment, data analysis skills, and facilitating meetings among the core teams to produce a schedule management plan. The schedule management plan is a work in progress throughout the project life cycle, as common revisions such as updated risks, value-added tasks, and preventive or corrective actions can change the plan. The schedule management plan, once complete, describes in detail the schedule for project activities, deliverables, and stakeholder communications. This plan forms a component of the overall project management plan and provides guidance and direction for other planning processes within project schedule management and other management plans.

The building blocks to a project schedule are *activities*, which are a distinct portion of work performed as a part of the Project Implementation Phase. Project activity characteristics include the following:

- A definitive start and end date
- Scope requirements
- Associated labor, costs, and other resources
- Measurable and controllable, and thus auditable
- Generally with defined input and output
- Clear person or group accountable and responsible
- High preference for a single point of accountability

Despite thorough efforts to plan the project schedule accurately, there are multiple factors that threaten the ability to complete projects on time. These factors include, among others, the following:

- Funding and cash flow (greater funding and cash flow enables more resources, which in turn allows a faster project completion)

- Resource availability (the capability, capacity, and availability of important resources impacts project progress)
- Activity type (some activities are resource intensive, whereas others require significant lead time)
- Dependencies (activities can be dependent on other activities, sometimes causing a knock-on delay and bottlenecks)
- Imposed deadline (some projects with regulatory ties cannot finish earlier than a preset date and time)
- Organizational culture (organizational culture can slow down or speed up project processes through decision-making practices and work values)

11.3 Planning, Estimating, Sequencing, and Validating Schedule

To implement project activities and tasks, project managers must achieve an acceptable degree of clarity as to what (scope), when (schedule), where (depend on project), who (resources – next chapter), and how. In predictive or traditional project management, the requirement for clarity is significantly higher than Agile methods. This is because in Agile approaches, schedule is often predetermined and fixed like a timebox. Within this timebox, the project team's goal is to find the optimal amount of scope that can be reasonably accomplished. Most of this section is dedicated to planning and estimating project schedule in the traditional project management approach, even though much of these concepts remain important for Agile approaches too.

Definition. Timebox is a technique for fixing time in which a task or activity must be accomplished.[1]

The main goal of schedule management during the Preparation Phase is to drive toward a high degree of clarity pertaining "when" should the implementation activities occur. There are four important subprocesses to achieve this.

1. Defining project activities
2. Estimating duration
3. Sequencing project activities
4. Validating project activities

[1] Williamson, B. & Wu, T., 2019. *The Sensible Guide to Key Terminologies in Project Management*, iExperi Press, Montclair, NJ. Glossary.

11.3.1 Defining Project Activities

Defining project activities is an important process of determining and documenting required actions to create project deliverables. From the work breakdown structure (WBS) as a part of scope management activities from Chapter 10, work packages are broken down into activities, which are considered the building blocks of a project schedule. Activities form the basis for scheduling, estimating, executing, monitoring, and controlling project progress. They describe the work to be performed in order to produce project deliverables. This process begins early in the Preparation Phase and is continuously revised and refined as new information about the project evolves.

Project managers review and consider inputs from the project charter, project management plan, organizational factors such as organizational culture, and information systems and apply their expert judgment, decomposition techniques, rolling wave planning, and facilitated meetings to define project activities. Teamwork is also important, as it improves idea generation, information gathering, and analysis for activity definition. Key outputs from this process include a detailed activity list with its attributes, the milestone, potential risks, and additional areas of concerns.

Definition. A work package is a small but measurable unit of work, typically at the lowest level of a WBS, in which there can be clearly associated schedule, deliverable or outcome, quality attributes, resources to perform the work, risks, and issues.[2]

Definition. *Rolling wave planning* is based on an Agile approach to product development and is an iterative process of planning at different levels of detail depending on the phase of the project life cycle. Work to be completed early on in the project is planned in detail, and the level of detail diminishes, as the work becomes longer term. As the project progresses, higher-level planning becomes more detailed, as the work to be completed grows closer.[3]

11.3.1.1 Key Considerations for Defining Project Activities

When defining project activities, it is important to focus on the project, both the project and the project *management* scope and requirements. This requires careful attention to detail, not omitting any key activities, and not adding any unnecessary activities.

---

[2] Williamson, B. & Wu, T., 2019. *The Sensible Guide to Key Terminologies in Project Management*, iExperi Press, Montclair, NJ. Glossary.

[3] Ibid.

The project manager must examine activities closely, because they may reveal additional activities or subactivities. In order to arrive at the most accurate process outputs, reviewing activities with subject matter experts, leveraging insights from similar projects completed in the past, and asking for assistance from experienced professionals can be useful. Once approval has been given on activity definitions, project managers must try to avoid adding further activities. If key activities are identified postapproval, then they should be put through a deliberate change management process.

11.3.2 Sequencing Project Activities

After defining the project activities, the next step is to sequence them according to relationships and dependencies. This process is performed repeatedly throughout the project life cycle. For complex projects, sequencing project activities is the process of determining and evaluating relationships among various project activities. Apart from the very first and last activity, each activity should have at least one documented predecessor and one successor. A predecessor activity is an activity that logically comes *before* a dependent activity in a schedule, whereas a successor activity is defined as a dependent activity that logically comes *after* another activity in a schedule.

To illustrate these relationships and dependencies, a network diagram is often used. While there are many forms of network diagrams, one of the most common is the activity on node diagram, also known as the precedence diagraming method (PDM). There are four types of relationships that exist between activities and tasks, as illustrated by the PDM in Figure 11.1.

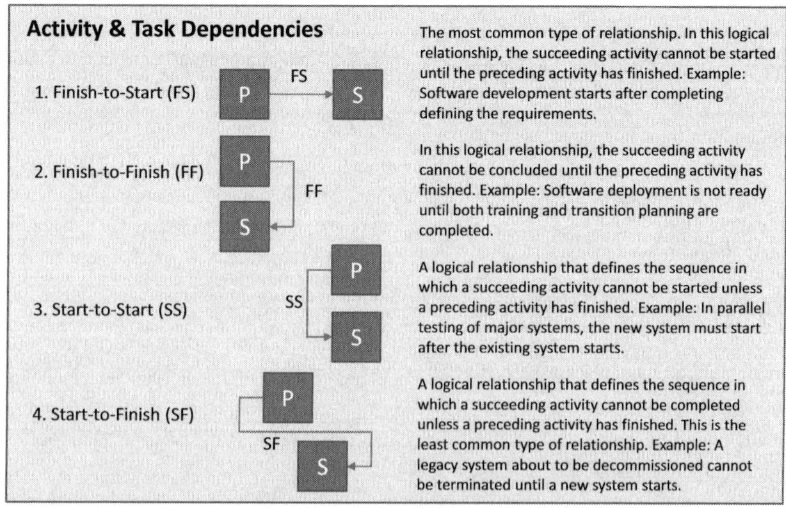

Figure 11.1 Activity and task dependencies. (Training and consulting content from PMO Advisory LLC. Reprinted with permission.)

The project management plan and project documents, such as the activity list and activity attributes, enterprise environmental factors, and organizational process assets, are valuable information for sequencing project activities. The PDM, dependency determination and integration, leads and lags, and project management information systems are useful process tools and techniques for determining the sequences. Upon completion of this exercise, the project activities would be properly sequenced, which can be represented in project schedule network diagrams.

11.3.3 Determining Dependencies

For project managers, determining activity or task dependencies takes place along two dimensions: internal versus external and mandatory versus discretionary: Mandatory dependencies, sometimes referred to as hard dependencies, are rigid sequence of activities as required by legal, regulatory, or process orders. For example, in baking cookies, one cannot start baking before shaping cookies. Discretionary dependencies, also referred to as soft logic, include sequences of activities based on common practices and subjective decision-making. For example, in making cookies, it's generally a good practice to mix the "wet" ingredients first such as butter and egg and add the "dry" ingredients such as chocolate or nuts later. Internal dependencies are activities that occur within the project team's locus of control, whereas external dependencies can be project or nonproject activities that lie outside of the project team's immediate control. Based on these dimensions, the following dependency combinations can occur:

Using a two-by-two matrix, these two dimensions form four possibilities:

- **Mandatory external.** Mandatory external dependencies are dictated by legal, contractual, or logical sequences outside of the project team's control. For example, in annual tax filing, an accounting team is waiting for updates from a software vendor that includes the latest IRS guideline before finalizing the tax preparation.
- **Mandatory internal.** Mandatory internal dependencies are dictated by legal, contractual, or logical sequences within the project team's control. For example, in annual tax filing, an accounting team is waiting for the financial team to issue the final profit and loss of the firm before completing the tax preparation.
- **Discretionary external.** Discretionary external dependencies are activity sequences (often nonproject activities) based on best practices and subjective decisions outside of the project team's control. For example, in annual tax filing, an internal accounting team is waiting for an external training session before starting the tax preparation.
- **Discretionary internal.** Discretionary internal dependencies are activity sequences (often project activities) based on best practices and subjective decisions within the project team's control. For example, in annual tax filing, an internal accounting team is waiting for a meeting with the finance team to discuss the profit and loss of the company before starting the tax preparation.

11.3.4 Estimating Project Activity Duration

Estimating project activity durations is the process of estimating the amount of time needed to complete individual activities with the estimated resources. This process occurs throughout the project life cycle. There are various factors that can influence the amount of time required to complete an activity, including the following:

- **Resources.** Increasing resources often reduces duration.
- **Technology.** Advanced technology for collaboration can reduce duration.
- **Project team motivation and team culture.** Harder working, motivated employees usually work faster and reduce duration.
- **Skill.** The more skilled the project team, the duration is likely to be shorter.

However, it is important to note that these factors usually do not affect duration along a linear model, nor do they guarantee a change in duration. Constraints, such as a preset activity time period (for example, a 3-day-long training workshop), or diminishing returns, often create a nonlinear increase or decrease in time.

For larger projects, estimating project activity duration is often a team effort in which the project manager engages the core project team. This is especially important when the project manager is not an expert in the project itself and thus acts more as a chief facilitator bring together various expertise. One of the first and most important things to discuss and agree is the unit of time used for project estimation and also the workday arrangements including holidays and organizational time offs (such as a plant shutdown). Based on the defined project activities created in the earlier process and also reviewing important project management artifacts such as the project management plan, the project team applies a combination of expert judgments and estimating techniques to estimate the duration required to complete the project activities.

There are a number of estimation techniques including analogous estimating, parametric estimating, three-point estimating, bottom-up estimating, data analysis, decision-making, and meetings. The three-point estimating technique often takes the form of the program evaluation review technique (PERT), which was developed in the 1950s for the Navy's Polaris system. This technique is designed for projects and activities with high uncertainty.

Definition. PERT formula for estimating activity duration is:
Expected Duration = (Most Likely + Least Likely + 4 * Most Likely)/6

The critical path method (CPM) is another popular approach to estimating duration. This approach is predominantly used at a project level, rather than activity level, and uses a single time estimate for each activity. Then, using a network diagram, one is able to find the longest sequence of activities by duration. The "critical path" is identified as the shortest time in which a project can be completed.

CPM is useful for defining the critical and noncritical activities in order to prevent project delays and process bottlenecks.

Despite the various techniques for estimating activity duration, the process is subjective, and so there are a number of challenges that project managers frequently encounter. These include the following:

- **The law of diminishing returns.** Effort is not sustainable, and productivity will decrease at a certain point due to constraints in a fixed factor
- **Parkinson's law.** Work expands to fill the time available for its completion[4] (bureaucracy)
- **Procrastination.** Stress, lack of motivation, and exhaustion of the project team or other stakeholders
- **Padding.** An unnecessary increase in estimated duration in order to increase confidence in the estimate
- **Excess optimism.** Decrease in estimate due to risk aversion or desire to please
- **Nature of activity.** Duration may be predetermined by the intrinsic nature of the activity
- **Nonlinear relationship with resources.** Additional resources may increase duration due to difficulty of coordination or required training

From the process of estimating project activity durations, the project will achieve the next level of clarity including project activities now with duration estimates and basis of estimates. This information should be used to update the project documents. If there are significant deviations from the duration estimate than the originally conceived duration, project managers should raise the finding with the project sponsor to reconfirm project support.

11.3.5 Developing and Validating the Integrated Project Schedule

After activities are defined, their sequence is determined, and activity durations are estimated, project managers can now develop the integrated project schedule. This process requires the analysis of project activity and it sequences, resource requirements (if available), durations, and time constraints to create an integrated project schedule for project implementation. The integrated project schedule will serve as a vital project management artifact for monitoring, reporting, and controlling. This is an iterative process requiring frequent and repeat interactions with costs, resources, risk, as well as influences from the other project knowledge domains.

Building on earlier schedule planning processes, the development of the schedule begins by defining start and finish dates for each activity and for the project as a whole. Analytical techniques, such as CPM, reveal the degree of flexibility and risk in

[4] Parkinson, CN., 1955. *Parkinson's Law*, The Economist, London.

the project schedule. As needed, project managers can apply schedule compression, contingency reserves and resource optimization refine the schedule.[5] Project manager and the team members then review their assigned activities within the schedule to identify conflicts and verify whether the schedule is realistic and acceptable. The validated and approved schedule model and milestones establish the schedule baseline for the project.[6]

Tool and technique. *Schedule* compression is an approach to project scheduling, whereby the duration is shortened, while not minimizing and/or reducing the scope or quality of the project in any way.[6] However, compression techniques often increase the overall project risk.

Typically, this technique involves *crashing*, where additional resources are added in order to complete activities more quickly, and/or *fast tracking*, where sequenced activities are performed simultaneously for a period of time. Fast tracking is only effective when activities can overlap along the critical path.

11.4 Controlling Schedule

During the Implementation Phase of the project, project managers must monitor the status of the project, including the project schedule, and adjust course as necessary to enable its timely completion. When changes occur, project managers should manage the schedule change deliberately. Once approved, a new schedule baseline should be created and used as the new reference point. In an Agile approach, there is an increased emphasis on specific time cycles and lessons learned.

Controlling the project schedule is an important part of the integrated change control process, and this process is performed throughout the project execution.

Tool and technique. Microsoft Project is a professional project management tool for Microsoft Windows Operating Systems. This is one of the most popular and arguably the most advanced desktop project management scheduling tools. All professional project managers should know how to create and manage a schedule using Microsoft Project. However, Microsoft Project can be expensive. An example of a networking diagram created automatically in Microsoft Project is depicted in Figure 11.2.

---

[5] Project Management Institute. 2011. *Practice Standard for Scheduling (2nd Ed.)*, Project Management Institute, Newtown Square, PA.

[6] Williamson, B. & Wu, T., 2019. *The Sensible Guide to Key Terminologies in Project Management*, iExperi Press, Montclair, NJ. Glossary.

Task Name	Duration	Start	Finish	Predecessors	Resource Names
Training Development	**37 days**	**Sun 3/1/20**	**Tue 4/21/20**		
A	5 days	Sun 3/1/20	Thu 3/5/20		JS,GB
B	3 days	Fri 3/6/20	Tue 3/10/20	2	GB
C	20 days	Wed 3/11/20	Tue 4/7/20	3	JS,GB,ET
D	15 days	Wed 3/11/20	Tue 3/31/20	3	JS
E	18 days	Wed 3/11/20	Fri 4/3/20	3	GB,ET
F	5 days	Wed 4/1/20	Tue 4/7/20	5	ET
G	10 days	Wed 4/8/20	Tue 4/21/20	4,7	JS,GB,ET

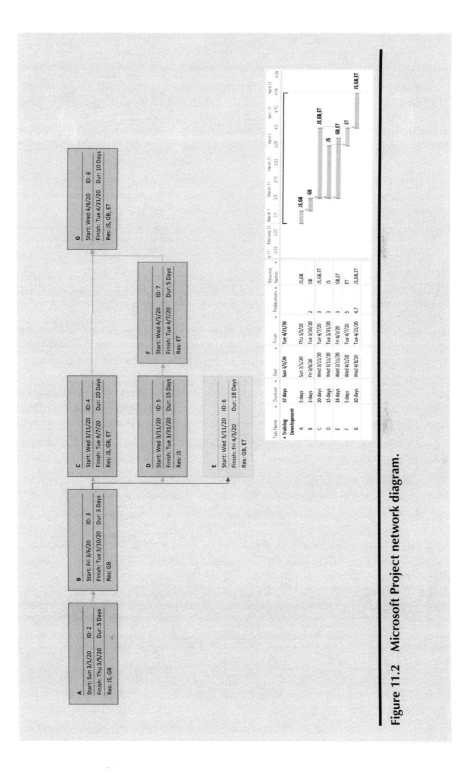

Figure 11.2 Microsoft Project network diagram.

As a substitute tool, Microsoft Excel and Google Sheets can substitute for project scheduling of simple projects. Its disadvantages are that it is very manual to use and has little ability to manage large projects with many tasks. Currently, there are over 300 online applications that cover some useful areas of project management including task and schedule management and collaboration.

Controlling project schedule can be challenging, because schedules are highly susceptible to change. Even with the best project managers, during complex and uncertain projects, it can become difficult to foresee all the activities during the Preparation Phase. Thus, the introduction of unplanned activities later on in the project life cycle can occur, and this is disruptive to activity sequences and may extend the critical path.

In some cases, customers or sponsors make changes to project scope, resulting in changes to the schedule. Project managers are also not experts in all components of projects, and so to obtain accurate estimates, they must work with subject matter experts on specific activity estimates. Even when consulting experts, all people are subject to bias, including optimism and pessimism. This can result in unrealistic estimates that cannot be achieved in the project execution. Delays can also be caused by inaccessibility to key resources on critical path activities. These resources can include project team members who are demotivated, slow workers, become sick, or even leave the organization. Any of these major disruptions can negatively impact schedule.

Project managers often have to analyze these risks upfront and plan accordingly. However, some risks (usually external risks) are beyond the prevention efforts of the project manager. There can be external events, often catastrophic, that negatively impact the project. For example, after the attack of 9/11, most of New York City Downtown was shut down for the following 3 months. This caused significant project delays, since most people left downtown hurriedly as part of an emergency evaluation. Natural disasters can result in the same delays. In order to mitigate these risks and proactively deal with the challenges of controlling project schedule, project managers should plan for contingencies for resources and backup valuable data regularly so that project information can be accessed remotely in the case of an environmental disaster or catastrophe.

To overcome positive and negative bias, project managers should inquire external experts or use objective tools, such as Monte Carlo simulations, to run risk analyzes. For changes that occur to the schedule, project managers should carefully follow the change management process for considering all changes, updating project documents, and communicating decisions to key stakeholders. Even with such proactive planning, there will occasionally be problems that are too large for project managers to control. In these cases, the project schedule will have to be expanded.

Therefore, it is important for project managers to work closely with all stakeholders. By developing relationships of high trust, project managers are able to work with executives to address these challenges.

11.5 Transitioning and/or Closing Schedule

As projects come to completion, it is important to transition or close the project schedule formally by obtaining agreements from the key project stakeholders such as customers and sponsors. Project managers should work with the project team to review progress and document key lessons learned pertaining to the project schedule. This is especially important for multiphase projects in which the project team is preparing and transitioning for the next phase of the project.

11.6 Project Management in Motion

This book is intended for a wide range of audiences and across various industries and functions. Therefore, to balance between competing demands of ease of use, comprehensiveness, modularity, flexibility, and upgradeability, the author has adopted a modular format to this book. This includes the relatively short chapters on key topics. In addition, the appendices include three vital sections:

- Appendix A contains a list of commonly used project management templates for both predictive and adaptive project management approaches.
- Appendix B contains an integrated case study based on a fictitious global company undertaking a number of projects and confronting various project challenges. The case maps closely to the chapters and sections, key concepts, and relevant tools of this book. The goal is to provide instructors, students, and practitioners with a realistic project example to practice and apply the key learnings in the book. For more information and additional case studies in which the author plans to create over time, including potential collaboration opportunities, visit www.optimizepm.com.
- Appendix C contains a glossary of selective terms, reprinting with permission from Dr. Te Wu and Mr. Brian Williamson's book titled "*The Sensible Guide to Key Terminologies in Project Management*", iExperi Press, Montclair, NJ.

Chapter 12

Resource Management – Defining the Resources Required to Tackle Projects

Summary

Projects can be resource intense, whether they are human resources, physical materials, remote technology such as cloud computing, or knowledge. From a system's perspective, resources are major inputs, and by managing them efficiently and using the best tools and techniques, project managers convert them to the desired outputs.

This chapter focuses on both the external and internal, people and nonpeople resources required to execute projects effectively. Capability, capacity, and resource planning are discussed, as well as the project manager's role in developing, managing, and controlling the project team.

This chapter focuses on addressing three important questions:

1. Why is resource management important?
2. What makes making people different – both positive and negative?
3. How to transform effectively resources to deliver project results?

12.1 What Is Resource Management?

Projects are often large and complex endeavors that require the sponsoring organization to invest a significant amount of resources in order to achieve the deliverables successfully. Project resources can be either human resources, physical assets, or even nonphysical assets. A single project can require a combination of these different forms of resources. Human resources usually include the people who form the project team, as well as their skills, competencies, and expertise. Physical assets include office space, equipment and machines, tools, and other physical resources such as concrete and steel for construction. Intellectual property, energy, and a virtual computing environment are examples of nonphysical resources.

Despite these differences, all resources share a few common traits:

- Resources are finite (some resources are even precious and rare)
- They are depletable
- They require funding for acquisition, maintenance, and/or use
- They can be contentious, either internally or externally, and some may have substitutes

In today's competitive world where organizations try to do more with less, organizational willingness to invest in resources is understandably limited. Therefore, given resources are finite, depletable, costly, and limited, it is the project manager's responsibility to manage the use of resources diligently to avoid waste.

There are four processes in the resource management knowledge domain, including the following:

1. Identifying project resource needs (Ideation and Initiation Phases)
2. Planning and estimating project resources (Preparation Phase)
3. Obtaining and working with resources (Implementation Phase)
4. Transitioning and/or closing key resources (Transition and/or Closure Phase)

12.1.1 Identifying Project Resource Needs

The discussion of resources required for implementation often arises early in the Ideation Phase. Generally, after the initial excitement of conceiving new ideas, reality sets in and questions of resources and funding start to enter the equation. As ideas harden into more concrete business plans and eventually into business cases, the clarity of resource needs becomes clearer. Project professionals may contribute to business case development, in particular estimating resources needs.

Depending on the organizational standard and processes and project complexity, these early estimations range from guesstimates to more accurate projection of resources. However, since ideas at this phase are still in the early development, the estimation of resources is limited by both the understanding of the ideas and the level of expert resources available to make firmer plans.

After the approval of business cases, projects now enter the Initiation Phase. During this phase, for both predictive and Agile projects, identifying and estimating resources is a major task for project managers and their team leads and subject matter experts (SMEs) in the development of the project charter and other project management planning documents.

12.1.2 Understanding Project Resources

Project resources refer to any inputs required for the implementation of projects. Project resources can be categorized into basic groups: human resources and non-human resources. Human resources are about people and their capabilities and skills. Nonhuman resources can have many subgroups; chief among them are physical resources such as machines, equipment, supplies, materials, and even the office space for the project team and intangible resources such as intellectual properties and licensing agreements.

Two valuable concepts in resource management are capability and capacity. Capability refers to a resource's ability, qualities, features, or power to perform required work. Capacity refers to a quantitative or qualitative amount of work that the resource is capable of performing. For example, in a simple project to paint a room, the capabilities include a painter who can paint the walls, a suitable paint that meets the customer's requirements, and tools that are used for painting. Capacity refers to the amount of paint, the number of brushes or other equipment, and the hours available from the painter.

One of the key challenges for most project-intense organizations is to find the optimal balance between the demand and the supply of resources. Therefore, project managers may need to conduct a capacity and capability analysis to evaluate the resources (human, finance, technology, machinery) of an organization required to implement projects, programs, or portfolios.[1] This is especially vital on larger projects in which organizational resources will be stretched severely.

12.2 Planning and Estimating Project Resources

Planning and estimating project resource is the process of identifying and defining the resources and estimating the quantity and quality of human resources and nonhuman resources for implementing the project, which includes both the project scope and the project *management* scope. Project managers, who typically work in an integrated fashion with scope definition and schedule development, examine

[1] Williamson, B. & Wu, T., 2019. *The Sensible Guide to Key Terminologies in Project Management*, iExperi Press, Montclair, NJ. Glossary.

and analyze project activities in order to determine the type of resources required for performing those activities. While most physical resources are relatively simple to plan and manage, human resources are often more complex. In particular, technical (hard) skills and behavioral (soft) skills are important considerations. Hard skills include project management capabilities (see Figure 12.1) and relevant subject matter expertise. Soft skills include the ability to work in teams and interpersonal skills. Resource planning should account for the availability of, and competition for, scarce resources, whether resources are acquired internally (within the organization) or externally (through the procurement processes).

Most of the project management artifacts created up to this phase of the project can serve as valuable inputs into the planning and estimating of project resources. These include project charter, components of the project management plan such as schedule, project documents, and organizational factors. Project managers and the core team apply tools and techniques such as expert judgment, organizational knowledge, facilitated meetings, and resource data analysis to estimate the quantity and quality of resources required to implement the project successfully.

One of the most popular tools for managing human resources is the Responsibility Assignment Matrix (RAM), of which the RACI (Responsible, Accountable, Consult, Inform) chart, depicted in Figure 12.2, is most common.

Definition. A *RAM* defines the connections between resource assignments, work packages, and activities. High-level RAMs match larger groups or teams to work breakdown structure (WBS) components, such as work packages. Low-level RAMs match individuals to specific work packages. The matrix format allows only a single entry for each work package, in order to eliminate confusion about assignments and responsibility.

Tool. A RACI chart is a special version of RAM that designates Responsible, Accountable, Consult, or Inform roles to team members. It is important that only one resource is accountable for each task.

- **Responsible.** Project team member who owns the delivering of this deliverable
- **Accountable.** Project leader (or team member) who reviews and approves the deliverable, thus owning the results
- **Consult.** Project team member who provides inputs, such as SMEs
- **Inform.** Project team member who must be informed of the outcome of the deliverable, but no input is required

Project managers should estimate the human resources needs and nonhuman resources such as the type and amount of materials, supplies, and the associated timing necessary to perform project work. On human resources, project managers should examine personnel needs, sources of human resources, the quality of

PROJECT MANAGER SKILLS

TECHNICAL SKILLS	BEHAVIORAL SKILLS
Analyze business requirements and create work breakdown structure	Manage stakeholder expectations
Estimate time, resources, and costs	Work with difficult people, especially in difficult situations
Create integrated project schedule	Select the right people for the project, especially in a diverse and international environment
Compress schedule when required	Create a high performance project team
Establish change management board	Deal with complicated situations, such as work schedule delays
Manage resources, including assigning people	Ability to handle stress, manage ambiguity and uncertainties, and lead teams
Institute project management processes to plan, monitor, and control projects	

Figure 12.1 Project manager skills. (Training and consulting content from PMO Advisory LLC. Reprinted with permission.)

	Person or Role				
Activity	Person 1	Person 2	Person 3	Person 4	Person 5
A	A	R	I	I	I
B	R	A	C	I	C
C	I	A	R	I	I
D	C	R	A	I	I
E	R	C	C	A	I
F	I	I	A	C	R
G	C	C	I	R	A

Figure 12.2 RACI (Responsible, Accountable, Consult, Inform).

resources pertaining to skills and experiences, and organizational policies with regard to hiring external contractors. In addition, there can be support considerations for people, such as information systems, physical space, or software, potential constraints on recruiting, scheduling, and releasing people, and the procurement of resources within the organization. Project managers should also consider creating resource calendars that provide information about the availability of human and physical resources.

Insights from the front line. There are a number of potential sources of human resources, each with their own challenges and complexities.

Organizational/functional/administrative units

Cost and internal transfer issues should be considered. Furthermore, in a matrix environment, there should be discussions on how to best manage these resources when there are multiple reporting lines.

Colocated team or virtual

Travel may be necessary when teams are geographically dispersed and contact sessions are required. Differing time zones and working hours may also need to be considered.

Skill and experience levels

Project managers must consider the required experience, as well as soft or hard skills for the project team. Typically, the more experienced resources require more funding.

Professional discipline

Depending on the type of work, range of responsibilities, and job titles, professional qualifications or certifications may be required. For example, a Project Management Professional (PMP®)–certified project manager might be required.

There are a number of tools and techniques for estimating resources that project managers and team can utilize. Chief among them includes expert judgment, data analysis, structured meetings, and estimating techniques such as bottom-up estimating, analogous estimating, and parametric estimating. Realistic resource estimates emerge from collaboration with multiple techniques and iteration and often in conjunction with schedule and cost estimation. Project team members often contribute their expertise in estimating resources required too.

Upon completion of this planning process, project managers should develop the resource management plan, team charter, and updates to the project documents. The resource management plan is the component of the project management plan that describes how resources are assigned to project activities to accomplish objectives. This document outlines the plan for human resources, including team members and other labor, as well as physical resources such as facilities, equipment, and supplies.

* "PMP" is a trademark of the Project Management Institute, Inc.

12.3 Obtaining and Working with Resources

After estimating the resources required, project managers must obtain the right resources for the project and also at the right time. These include project team members, equipment, supplies, facilities, and other resources required to achieve project outcomes. Since resources can be expensive, have a limited useful life (perishable products or even technology), and can take a significant lead time to acquire, it is important to plan the schedule of acquiring resources accurately. This can be most important for human resources, where considerations include the timing of when project team members should be onboarded, their enculturation on the team, motivating team members, providing a purpose to team members, when and how to use rewards and incentives, quality assurance and oversight, and releasing project team members from the project.

While internal resources, which are assigned by functional or resource managers, are often more controllable, external resources acquired through a procurement process can be more of a challenge. These resources can be constrained by contractual agreements or external lines of reporting. Insufficient or inappropriate resources can compromise project quality and completion, client satisfaction, and project costs, so it is important that the project manager uses influence and negotiation to ensure the timely and appropriate acquisition of external resources. It should be noted that acquiring internal resources may not be simple, as there can be internal politics, training required for the internal resources before they become contributors, and other personnel factors.

The process of obtaining resources can be performed throughout the project as resources are invited to join the project and released when they complete their work. Project managers utilize important inputs from the available project artifacts, including project management plan components, organizational resource planning, and other project documents. When working with people, project managers may work in conjunction with human resources specialists to determine the optimal balance of technical and soft skills and physical and virtual teams. There are many outcomes from the acquire resources process, which include resource allocations, team assignments, resource calendars, project management plan updates, change requests, and updates to other project documents such as WBS and integrated project schedules. There are also a number of project documents created and leveraged by project managers, especially for large and complex projects. Examples include the following:

- Project schedule outlines detailed activities and their timing.
- Resource calendar highlights when resources are needed and should be available for the project.
- Resource requirements identify the basic requirements of the resources.
- Stakeholder register lists the human resources (stakeholders) on the project, as well as their needs and expectations.

Some organizations also have clear policies and guidelines with regard to the acquisition of people. These policies and lessons learned can be important information for the project manager.

Throughout the Project Implementation Phase, project managers should closely monitor the resources and work with them to make adjustments as required to control its utilization. Since resources are often intricately linked with schedule and cost, change in any of these areas should trigger an analysis of the other knowledge domains.

12.3.1 Developing Project Teams

Human resources require a special focus in order to maximize their effectiveness and value to the project team. After acquiring the people resources, the project manager must develop the project team. The chief goal is to create a capable team, perhaps even a high-performance team, by improving team members' competencies, enabling quality interactions among them, and developing a positive team environment. The key benefit of this process is that it enhances overall project performance by creating excitement and a sense of belonging among team members, reducing attribution, improving motivation, and creating an environment of learning and nurturing. As one of their primary responsibilities, the project manager must ensure the development of a team environment and culture that promotes trust, respect, and collaboration, empowers decision-making, enables effective communication, and leverages cultural diversity and knowledge sharing. The process is performed throughout the project.

To create a high-performance team, the project manager develops, communicates, trains, and reinforces a basic set of team operating principles that are outlined in the team charter. This includes meetings and how they will be conducted, decision-making guidelines, clarification on the escalation of issues, risks, and changes, establishing how work gets done, including reporting and team values. The resource management plan may be revised based on interactions with other project documents or plans. The plan resource management process is performed once during Project Preparation Phase and continues to serve as a robust yet flexible reference point for resource management throughout the project.

Project teams move through a number of sequential stages, although they may occasionally skip a state or regress to a previous state. One of the most well-known models for team development is the Tuckman Ladder[2], which includes five sequential stages:

1. **Forming.** Team members become acquainted with each other; they tend to operate independently as they learn about others on the project team

[2] Tuckman, B., 1965. Developmental sequence in small groups. *Psychological Bulletin*. 63(6): 384–399.

2. **Storming.** Team members start to work together; there can be competition for roles, and tension and conflict due to differences in ideas
3. **Norming.** Team members begin to adjust their behaviors to conform to the team's norm; they learn to trust one another
4. **Performing.** Team members are able to work together, and the primary focus is on performance
5. **Adjourning.** Team members complete project activities and move on to the next assignment

A good practice to bring the team together is the establishment of ground rules, which are the basic rules and expectations for acceptable team behavior.[3] Project managers can use the exercise to establish the ground rule as a team building exercise. This way, project team members are involved and informed of the expected behavior. This can also prevent and minimize potential misunderstandings and conflicts.

12.4 Leading Teams and Controlling Resources

During the Project Implementation Phase, project managers need to lead, manage, and control project team member performance and intervene as necessary to provide feedback or resolve issues. This includes encouraging and motivating team members and optimizing people and teams to achieve project results. Project managers should assign people to specific activities and levels of responsibilities and monitor their willingness and ability to perform in their roles. If required and when feasible, project managers can leverage projects as developmental activities for team members. Of particular importance is the project manager's ability to facilitate teamwork, communication, conflict resolution, and leadership in order to achieve a high-performing team.

Resource assignments should be centralized in a resource plan and analyzed for potential issues, such as overloading. During implementation, project managers will leverage the project management artifacts created during the earlier life cycle phases and apply their knowledge, experience, and professional judgment to maintain or adjust actions. The most notable outcome of strong leadership is high-performing teams that effectively and efficiently executed the project plans and successfully delivered results.

See Chapter 20 on Working with People for more information on leading and motivating team members.

---

[3] Williamson, B. & Wu, T., 2019. *The Sensible Guide to Key Terminologies in Project Management*, iExperi Press, Montclair, NJ. Glossary.

12.4.1 Controlling Resources

On resources, in addition to leading people, project managers are also responsible to manage project resources and oversee their utilization so they are performing as planned. When necessary, project managers should be taking corrective actions to maintain project progress and performance baseline. Corrective actions may be necessary regardless of the project approach. In predictive approach, estimating project resources (as well as duration and cost) are not always accurate, and so actual resource utilization must be monitored and controlled throughout the project. In Agile or adaptive approach, resources are more likely to be fixed. Thus, project managers or equivalent many need to modify scope to best fit the Agile project in an iteration with fixed resources and duration.

One of the common challenges in controlling resources is the mismatch between the demand required by the project and the supply of resources. For human resources in particular, this can create an unhealthy project environment when resources are overloaded or assigned without being utilized. Project managers can manage resource overload by identifying the resources that are overloaded and either obtaining more resources or distributing the work. Specific methods for resolving overloads can include the reordering of project activities, acquiring or transitioning existing resources to assist, reducing activity and project scope, and extending or adjusting the project schedule. When the resource mismatch is significant, it is important that the project manager works with sponsors to address the problems early on.

Definition. *Resource leveling* is a technique for resolving overload in which start and finish dates are adjusted based upon resource constraints. The goal of this technique is to balance the demand for resources with the available supply. Project managers can also shift resources from noncritical path activities to work on the critical path activities. While this may work in some situations, it may not work in others, because the skills required for certain activities could be rare and difficult to train. Resource leveling on large projects can be particularly difficult, so project managers should work with their sponsor, PMO managers, and program and portfolio managers for large-scale leveling. Any changes in resources should also be managed formally as part of the change control process.

12.5 Transitioning and/or Closing Key Resources

Resource transition can occur throughout the project. As team members complete their project activities, they should be released from the project so the project no longer needs to pay for these resources. At the project closure, all project resources should be transitioned off the project. This would enable the project manager to close the project fully.

For nonpeople resources, most of the depletable resources should ideally be spent. The remaining resources can be transitioned to operations or the next phase of the project. In some cases, unused resources can be returned to the supplier.

For human resources, it is important for the project manager to work with human resources or other organizational leaders to shift people to their next assignments. Whenever possible, project manager should engage their key resources in the completion of postproject evaluation as well as updating key project documents. In some organizations, project managers may be responsible for reporting formally or informally personnel performance and contribution. It is important to invite resources to project celebrations and to recognize their contributions.

12.6 Project Management in Motion

This book is intended for a wide range of audiences and across various industries and functions. Therefore, to balance between competing demands of ease of use, comprehensiveness, modularity, flexibility, and upgradeability, the author has adopted a modular format to this book. This includes the relatively short chapters on key topics. In addition, the appendices include three vital sections:

- Appendix A contains a list of commonly used project management templates for both predictive and adaptive project management approaches.
- Appendix B contains an integrated case study based on a fictitious global company undertaking a number of projects and confronting various project challenges. The case maps closely to the chapters and sections, key concepts, and relevant tools of this book. The goal is to provide instructors, students, and practitioners with a realistic project example to practice and apply the key learnings in the book. For more information and additional case studies in which the author plans to create over time, including potential collaboration opportunities, visit www.optimizepm.com.
- Appendix C contains a glossary of selective terms, reprinting with permission from Dr. Te Wu and Mr. Brian Williamson's book titled "*The Sensible Guide to Key Terminologies in Project Management*", iExperi Press, Montclair, NJ.

Chapter 13

Cost Management – How to Develop and Manage Budgets

Summary

When the ancient Egyptians built the pyramids or when the Chinese constructed the Great Walls, the cost was often a postscript. After all, what can be more important to pharaohs who seek a never-ending life or to the Chinese of mitigating existential threats of invaders? In more recent times, the Manhattan Project began modestly, but as World War II continued, the project's importance became paramount and cost was no obstacle. The United States spent $2 billion, which in 2016 dollars would be around $27 billion according to Wikipedia. Few projects today, government or private, have the luxury of the seemingly unlimited budget. Rarely do project managers complain of too much budget and not sure how to spend it. In business organizations, it is far more common that projects are underfunded, as management urges their project teams to "do more with less".

In this chapter, we discuss the complexity and challenges of planning, estimating, and controlling project costs, as well as steps for determining the project budget. Tools and techniques for cost estimation, budgets, and cost management are explained, and common issues and consideration for project managers are emphasized.

This chapter focuses its attention on addressing these three essential questions pertaining to cost management:

1. Why is project cost management important?
2. What can go wrong?
3. How to more effectively manage project costs?

13.1 Effectively Plan Project Cost

Projects can be costly, and when project costs are not properly managed, the negative consequences can become large. According to the PMI Pulse of the Profession[1] report, as much as $122 million is wasted for every $1 billion spent, primarily due to poor project performance. Incorrect resource estimates, scheduling delays, scope creep, misaligned expectations, and other unforeseen factors can lead to increased costs and the failures to stay within project budget. Possibly the most relevant example of this is the Sydney Opera House. Its construction was subject to significant delays, and at its completion, the project had cost nearly ten times the allocated budget.[2] For this reason, one of the primary responsibilities of the project manager is to plan, estimate, budget, manage, and control project costs in order to achieve project deliverables within the approved budget.

There are four processes in the cost management knowledge domain, namely:

1. Proposing initial project costs (Ideation and Initiation Phases)
2. Planning and estimating project costs and budget (Preparation Phase)
3. Managing and controlling project costs (Implementation Phase)
4. Transitioning and/or closing project costs (Transition and/or Closure Phase)

13.2 Proposing Initial Project Costs

During the Project Ideation Phase, once ideas move behind the initial stage of consideration, question of costs arises. After all, one major component of business cases of most initiatives is the financial feasibility analysis in which costs take center stage. Similar to schedule and resources, cost estimations during Ideation Phase are based on limited available information and limited expert resources. Initial project cost estimates are usually developed by the person or group who developed the idea.

---

[1] Project Management Institute. 2016. *The High Cost of Low Performance: Pulse of the Profession.*

[2] Moore, S. W., 2009. The Sydney Opera House and project management. *Strategic PPM.* Retrieved from https://strategicppm.wordpress.com/2009/09/25/the-sydney-opera-house-and-project-management/ on November 10, 2019.

Thus, the costs are likely to be inexact during the Ideation Phase. Throughout this phase, as ideas solidify into concrete initiatives and later into projects, cost estimates also become firmer.

After the approval of the business case, the project enters the Initiation Phase. Here, project managers examine the business cases, and working with a core team starts to develop the next level of project cost with the intention of formalizing various costs into a project budget. A preliminary budget is either embedded in project charter or done as a separate analysis, and this budget should include costs for both the project implementation and the project management work required to deliver the project successfully. Upon their approval, projects move into the Preparation Phase.

13.3 Planning and Estimating Project Costs and Budget

Planning and estimating projects and budget occur in the Project Preparation Phase, and it is the process of developing detailed costs and assembling them into a project budget. For Agile projects, the cost planning and budget estimation are similar to traditional project methods. The primary difference is that parts of the project costs may be predetermined by fixing the duration of each iteration and resources available. However, Agile projects often do not know the number of iterations at the beginning of the project, even though there can be good guesstimates. Plus, the scope is typically not as fixed and rigid. Therefore, the flexibility of the project scope and the number of iterations provides a degree of freedom for project managers to maneuver and optimize for the project outcome successfully.

13.3.1 Planning Cost Management

For larger and complex projects, preparing a detailed cost required management planning. The process of planning cost management defines how the project costs will be estimated, budgeted, managed, monitored, and controlled. This process is performed once during the Preparation Phase and can serve as a robust yet flexible reference point for tracking and controlling cost performance throughout the project.

Project managers spearhead the development of the cost management plan by utilizing the previously created project management artifacts such as the project charter, project management plan, organizational factors and budget guidelines, and other project documents. Project managers also work with sponsors and organizational leaders to define the costs that are a part of the project budget versus the organizational budget. For larger endeavors, project managers may also need to work with program and portfolio managers to determine the project cost and budgetary processes.

With that understanding, project managers apply their expert judgment, analyze data, and facilitate meetings to develop the cost management plan. This is a document that describes in detail how costs will be planned, structured, and controlled during the project, as well as the supporting tools, techniques, and processes for cost management. The cost management plan forms part of the project management plan and may vary in its level of detail and formality depending on the complexity of the project. It provides the framework for efficient performance and coordination during monitoring and controlling, may be revised based on interactions with other management plans, and provides guidance for other planning processes under project cost management.

Insights from the front line. Some organizations take a comprehensive view on costs and capture employee compensation, benefits, office space, and all other costs as part of the project budget. In other organizations, only additional expenses are included as a part of the project budget, while employee remuneration is excluded. In the middle, there can be a range of rules and guidelines depending on the project and the organization.

As a new project manager in your organization, make sure you inquire and determine the organization's rules and process for project cost inclusion.

13.3.1.1 Cost Management Plan Components

A cost management plan serves to document project costs clearly in order to enable timely communication and accurate representation and to link the most current project cost with the business case estimation. These connections can be vital to engender continual support of the project. The plan also provides details to the project team, which is important on larger projects where there are multiple stakeholders. Finally, a cost management plan ensures that important stakeholders are aware of all costs (as well as schedule, resources, and scope) to provide a high degree of transparency in the management of the project. Components of the cost management plan are outlined in Table 13.1.

13.3.2 Estimating Project Cost

Costs are closely linked to project scope, schedule, resource planning, and quality. In estimating project cost, project managers approximate the cost of project resources needed to complete the project successfully. This is an iterative process to determine the monetary resources required for each project activity. In this process, human and physical resources, schedule, quality, and risk are continuously traded off against each other in order to maintain the project budget.

To estimate project costs, project managers utilized the available project artifacts such as the project management plan and project documents from previous similar

Table 13.1 Cost Management Plan Components

Units of measure	Metrics for currency or other resource units related to costs, such as time and materials measures
Levels of precision	Degree of rounding up or down for estimates; depends on scale and scope of project
Levels of accuracy	Acceptable ranges (e.g., +/–10%) for realistic cost estimates and cost contingencies
Organizational procedure links	Framework for interfacing between work breakdown structure control account and accounting system
Control thresholds	Degree of flexibility in cost baseline before variance requires action; typically expressed as a percentage deviation
Type of cost	The six common types of project cost include fixed, variable, direct, indirect, sunk, and contingency

projects including project archives and the lessons learned register and apply estimating technique and sound professional judgment to more accurately estimate project cost. The accuracy of cost estimates is highly reliant on collaborating with experts, multiple iterations, and reviews and applying multiple techniques. Once cost estimates have been determined, they should be expressed in a preagreed currency unit. This is especially important on international projects where there are multiple currencies.

Definition. The following are commonly used terms in project cost estimation:

An *estimate* is a quantified assessment of the likely amount based on science or experience. It should always include an indication of accuracy and precision.

Accuracy describes the closeness of the estimate to the actual cost.

Precision describes the closeness of multiple estimates to each other, but not necessarily to the actual cost.

A *reserve* is a provision in the project management plan to mitigate cost and/or schedule risk. Two common types of reserve include the *management reserve* and *contingency reserve*. The project sponsor and senior management hold the management reserve. The amount of this reserve corresponds to *unknown* risks that may affect the project. The project manager typically holds the contingency reserve. The amount of this reserve corresponds to *known* project risks.

Estimating project costs can subject project managers to a number of ethical and professional considerations. For example, there may be pressure to suppress certain costs or the entire proposal in order to win. Practitioners and especially members of the Project Management Institute (PMI) should review the PMI Code

Table 13.2 Factors Influencing Cost

Factors	*Description*
Type of cost	There are many types of costs incurred during a project. • Fixed versus Variable • Direct versus Indirect • Recurring versus Nonrecurring • Regular versus Expedited • Internal versus External • Lease/Buy versus Make • Labor versus Materials • Estimate, budget, and reserve
Accuracy, timing and method	Estimates should be done at three levels. All estimates should be verified with subject matter experts who have experience with similar projects. 1. *Order of Magnitude* (also known as indicative) Timing: estimation during the business case Accuracy range: +50% to –50% 2. High-level (or parametric) Timing: estimation during the Preparation Phase Accuracy range: +30% to –20% 3. *Detailed* (also known as definitive) Timing: estimation during the beginning of the Implementation & Quality Assurance Phase Accuracy range: +15% to –5%.

of Ethics and Professional Conduct.[3] Some important principles to serve as guidelines in project estimation include the following:

- Never lie to yourself and others. Clearly document project assumptions.
- Acknowledge that there are situations in which only selective information can be shared. However, project managers should document supporting information, even though they may only be shared internally.
- Understand the estimation techniques and their limitations. This is especially important for larger projects, as there can be many rounds of estimates. Understanding the techniques can provide longitudinal analysis of project costs and its evolution.

There are a number of other important factors to consider when estimating project cost. Table 13.2 shows two common factors: type of cost and its accuracy, timing, and method.

[3] PMI. 2016. *Code of Ethics and Professional Conduct.*

There are also many issues and concerns that can influence cost estimates. While there is no definitive list of potential issues in estimating project cost, there are some commonly occurring issues that should be considered during estimation. These include the following:

- **Information asymmetry.** This occurs when the project manager does not have all the information, possibly because vendors do not readily reveal the true costs. In extreme situations, this can even be due to deception and fraud.
- **Inadequate planning.** For large and complex projects, it is difficult to have sufficient details early enough in the project to produce accurate estimates. As the project evolves, through progressive elaboration, project managers can obtain realistic estimates.
- **Disruption in the market.** On resource-intensive projects, material costs can be susceptible to market fluctuations.
- **Exchange rate.** For projects across borders, exchange rates may need to be managed in order to steady the cost of materials and other resources.
- **Time value of money.** For long-term projects, the financial analysis should consider the duration, and the associated inflation over this period when allocating budget.

13.3.3 Determining Budget and Controlling Cost

Based on the estimated project cost, and often through multiple iterations of review and validation by subject matter experts, the project sponsor, and other executives, the project manager is ready to determine the project budget. Determining budget is the process of combining and totaling the estimated costs of individual activities or work packages in order to create an authorized cost baseline for the entire project. Project managers often work closely with experts and project sponsor in the assembly of the project budget at the beginning of the project. For larger projects, this process can be repeated at predetermined points throughout the project life cycle or when there are major changes that require revisiting the project budget. This process is similar for both adaptive and predictive projects.

Typically, there are four important steps to assembling the project budget.

1. The first is to *aggregate costs* from the activity cost estimates. This step involves the addition of all project cost estimates at an activity level in order to arrive at a consolidated budget. It is important to include all costs in this step, including project and project management costs.
2. The second step is to *review risks and analyze contingency and management reserves.* Contingency reserves prepare for risks that are known. These reserves can be developed through risk identification. See Chapter 15 on Risk

Management. Management reserves prepare for risks that are yet unknown, also referred to as "unknown unknowns". Project sponsors should consider the likelihood of unknown unknowns occurring and plan to mitigate their impact.

3. The third and last step is to *determine the accounting approach and funding mechanism required,* including funding disbursement. This step will be important for determining project cash flow. For international projects, project managers should also prepare for exchange rate fluctuations affecting the budget.
4. The fourth and final step is to *review the draft budget* with the project sponsors, core team members, and other key stakeholders to ensure continual support. If the draft budget does not achieve funding support, project managers should work with the sponsor to determine the path forward. This can include the termination of the project, as it may no longer make financial sense, revert back to the Initiation Phase or even Ideation Phase for a complete revamping of the project, and reexamine project activities and duration with the core team and customer to descope and reprioritize certain requirements. Under certain business situations, customers may ask project managers to expedite projects at an increased budget.

13.4 Managing and Controlling Project Costs

In the Implementation Phase of a project, the project manager should regularly monitor the project budget, forecast the spending, managing cash flow, and ensure that the project spending lies within the approved budget. To implement cost controls, project managers often establish milestones as measuring points to determine the actual versus planned project spending, including cash flow projections versus expected funding. Project managers should gather and analyze project performance data and evaluate them against the key project management artifacts such as project funding requirements, cash flow or funding disbursements, project management plan components, and other project documents to manage project budget successfully.

13.4.1 Earned Value Management System

Within the data analysis technique, an important method for managing and controlling cost is the "earned value management system" or EVM. This is a project management technique for measuring project performance objectively. EVM has only a few essential concepts that must be understood, such as "value" being the worth of progress and accomplishment, "planned goals and objectives" being the *target* goals and objectives, and "actual" being the measurement of what has been achieved. The EVM method primarily measures two dimensions, namely cost and time. Using basic arithmetic, EVM provides a robust set of project management performance measures and calculations, such as the Cost Performance Index (CPI), and Schedule Performance Index (SPI). However, the concept of "value", or what constitutes value, can be challenging to define objectively on projects. Using an analysis technique such as EVM, project managers can produce work performance reports, cost forecasts, change requests, and important updates to project documents.

EXAMPLE OF EARNED VALUE MANAGEMENT

You are the project manager for an upcoming annual conference. The project has a budget of $50,000, and the timeframe is 100 days. The planned "burn rate" is $500 per day on average. By day 10, the project is clearly progressing faster than expected. The number of completed activities is 10% further than planned, and you have actually only spent about 10% of the anticipated budget for this period. You would like to calculate whether the project is running under or over budget.

Project figures:

- Budget at completion in $ = $50,000
- Budget at completion in days = 100 days

Based on the above information, the earned value calculations are shown in Table 13.3.

Table 13.3 Earned Value Management Example

Term	*Acronym*	*Formula*	*Example**
Expected monetary value	EMV	Σ (prob.* impact)	$50,000
Budget at completion, $	BAC, $		$50,000
Budget at completion, days	BAC, days		100 days
Planned value	PV		$5,000
Earned value	EV		$5,500
Actual cost	AC		$4,500
Cost variance	CV	EV – AC	$1,000
Schedule variance	SV	EV – PV	$500

(Continued)

Table 13.3 (*Continued*) Earned Value Management Example

Term	*Acronym*	*Formula*	*Example**
Cost performance index	CPI (>1 favorable)	EV/AC	1.22
Schedule performance index	SPI (>1 favorable)	EV/PV	1.10
Estimate at completion	EAC	BAC / CPI or AC+ETC	$40,909
Estimate to complete	ETC	EAC – AC	$36,409
Variance at completion	VAC	BAC – EAC	$9,091
To-complete performance index**	TCPI (<1 favorable)	A. (BAC – EV) / (BAC – AC) or B. (BAC – EV) / (EAC – AC)	0.98
Time estimate at complete	TEAC	BAC Days / SPI	90.9
Time variance at complete	TVAC	BAC Days – TEAC	9.1

Notes:

* The column provides an example of earned value management.

** TCPI has two formulas. Use "A" when the project is under budget and "B" when the project is over budget. In the example above, the project is under budget because the CPI is greater than 1.

13.5 Transitioning and/or Closing Project Costs

As projects complete their implementations and entering the final phase of the project, the Closure Phase, project managers should review and analyze project costs and update the project budget and refine the definitive cost estimates. There are two important closure activities pertaining cost: (1) closing the project budget and (2) capture key lessons learned.

On closing the project budget, project managers should update all the project cost items and determine the actual spent versus the project budget. It is especially important to account for all the cost-related changes. Project managers should update key project management artifacts with the latest cost information. In addition, project managers should document the reasons and rationale for variances between the actual and the budget and submit that as a part of the final project closeout report. If there are remaining unused funds, project managers should work with project sponsors to determine its management, such as return the unused fund back to the sponsoring organization.

Beyond documenting the rationale for cost variances, project managers should also use this opportunity to work with the core team members to capture the key lessons learned related to costs from this project. This will ensure future projects can leverage both the successes and improvement opportunities.

13.6 Project Management in Motion

This book is intended for a wide range of audiences and across various industries and functions. Therefore, to balance between competing demands of ease of use, comprehensiveness, modularity, flexibility, and upgradeability, the author has adopted a modular format to this book. This includes the relatively short chapters on key topics. In addition, the appendices include three vital sections:

- Appendix A contains a list of commonly used project management templates for both predictive and adaptive project management approaches.
- Appendix B contains an integrated case study based on a fictitious global company undertaking a number of projects and confronting various project challenges. The case maps closely to the chapters and sections, key concepts, and relevant tools of this book. The goal is to provide instructors, students, and practitioners with a realistic project example to practice and apply the key learnings in the book. For more information and additional case studies in which the author plans to create over time, including potential collaboration opportunities, visit www.optimizepm.com.
- Appendix C contains a glossary of selective terms, reprinting with permission from Dr. Te Wu and Mr. Brian Williamson's book titled "*The Sensible Guide to Key Terminologies in Project Management*", iExperi Press, Montclair, NJ.

Chapter 14

Communication Management – Ensuring the Full Handshake

Summary

This chapter focuses on the importance of communication on projects and discusses a variety of methods to plan, manage, and monitor the effective and efficient flow of information among project stakeholders. Even though project managers often spend 80%–90% or even more of their time on communication-related activities, communication challenges and issues are one of the most often cited problems.

To strive toward better communication management, this chapter presents a systematic view of planning communication, implementing the communication plan, monitoring the results of communication, and adjusting or controlling as required to improve communication.

This chapter addresses the three most important questions in project communication:

1. Why is project communication important?
2. How to plan and execute communication effectively?
3. How communication and stakeholder management work together to enable better project execution?

14.1 Importance of Project Communication

Communication is the exchange of information via various mediums (e.g., face to face, phone, e-mail, and chat) and in differing formats (e.g., written, verbal, formal, and informal). Action-oriented communication is an example of project communication and includes the communication of work authorization and approvals (e.g., steering committee approval of the next iteration), project decisions (e.g., to reject or approve change), and project tasks (e.g., project managers assign work to project team members). Another example of project communication is documentation, which includes project management documents, such as the project charter, project management plan components (issue log, risk register, communication plan), stakeholder analysis, contracts, lessons learned, and project documents, including project requirements, design, quality procedures and standards, training materials, and operational procedures.

Effective communication is essential for achieving a mutual understanding of project goals and objectives, stakeholder interactions and contributions, minimizing costly misunderstandings, and developing relationships with trust and commitment across diverse cultures, interests, and expertise. Timely, productive, and direct communication between stakeholders is particularly beneficial in projects operating with high levels of change and uncertainty, because communication becomes a tool for promoting agility.[1] Effective communication thus increases the likelihood that organizations will achieve project goals and deliver projects on time and within budget.[2]

Measured by intensity and frequency, project communication processes are arguably the most frequently applied and used processes in project management. The responsibility of managing even a medium-sized project with over ten stakeholders requires frequent communication across various channels. This is in order to evaluate the project, identify and coordinate work tasks, tackle issues, manage change, engage risks, and work with executives and other important stakeholders to maintain support and resources and ultimately achieve project success.

There is a simple formula to calculate the number of communication channels surrounding a project. Figure 14.1 illustrates this point by using the formula, Communication Channel = n * (n – 1)/2, where n = # of people. As observed in this figure, there is a steep increase in the number of communication channels relative to the number of stakeholders until the point of about 200 stakeholders. The ratio of communication channels to stakeholders thus decreases as the number of stakeholders increases.

---

1 Project Management Institute. 2017. *A Guide to the Project Management Body of Knowledge (PMBOK® Guide - 6th Ed.)*, Project Management Institute, Newtown Square, PA.

2 PMI. 2013. *Pulse of the Profession™ In-Depth Report: The Essential Role of Communications*, Project Management Institute, Newtown Square, PA.

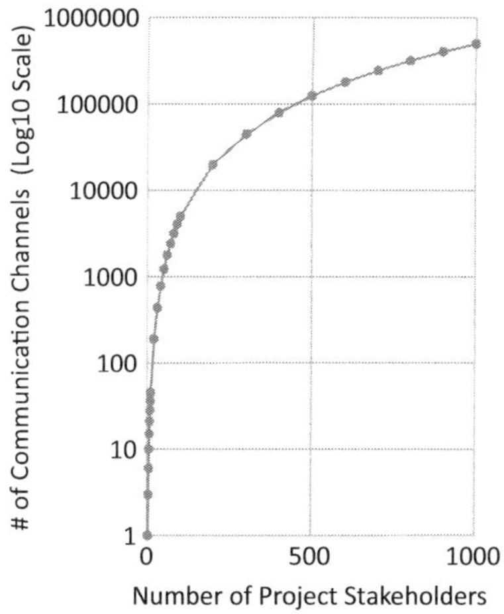

Figure 14.1 Stakeholder engagement by number of channels.

Project communications management includes processes for developing and implementing strategies for effective information exchange and stakeholder engagement and building relationships between internal, external, and culturally diverse stakeholders. The three processes for communication management are as follows:

- **Developing communication and information management plans** (Ideation, Initiation, and Preparation Phases)
- **Managing and monitoring communication and information exchanges** (Implementation Phase)
- **Transitioning and/or closing communication and information management** (Transition and/or Closure Phase)

14.2 Developing Communication and Information Management Plans

Project information is a valuable asset that must be managed through carefully planned communication. During the early life cycle of projects, information is created, refined, manipulated or adjusted, disseminated, and aggregated. Selective and sensitive information may be destroyed after the project's completion. Developing communication and information management plans is the process of creating an optimal project communication and information management approach in order to ensure that appropriate information is properly managed, transmitted, and received and used by each stakeholder or group. The ultimate goal is to ensure that the right information reaches the right stakeholders at the right time. Through effective management of information and communication, project stakeholders can access the right information efficiently. An additional goal is to transform *information* into *knowledge* that can benefit the entire project.

The emphasis on treating information as a strategic asset begins early in the project life cycle and can start as early as the Ideation Phase when key stakeholders are identified. In the Initiation Phase, communication planning and information management can be an important analysis and input into the project charter. This process of planning communication and managing information often reach the highest intensity in the Project Preparation Phase in the development of project communication plan, information, or knowledge management plan. Often conducted simultaneously with identifying and analyzing stakeholders, these plans serve as important components of the project management plan. Even though these plans are created mainly in the Preparation Phase, they should be reviewed regularly throughout the project life cycle especially when there are major changes to the project. Depending on the nature of the project and the sensitivity of information, security and access can be important factors for consideration during this process.

From a process perspective, project managers leverage the available project management artifacts such as the project charter, project management plan components, organization's existing communication structure, and other project documents with a special focus on stakeholder-related artifacts such as stakeholder register and engagement plans. While all project stakeholders require communication about the project, the type of information and the channels through which this information is communicated often differ according to the stakeholder. In order to recognize these diverse communication needs, a number of tools and techniques are applied to this process. Project managers, working closely with the core project team, apply their expert judgment, leverage existing or creating new communication models and methods, employ interpersonal and team skills, and use data representation and meetings to develop communication plan and information management plans. On larger projects, project managers may need to conduct special communication requirements analysis, which is used to determine stakeholders' informational needs. Using sources such as the stakeholder register, stakeholder engagement plan, organizational charts, team logistics and locations, and a number of other sources, requirements are determined by combining the information type and form (see Figure 14.2) with the perceived value of information to stakeholders.

Communication technology is another important tool of consideration, which is used for information and knowledge exchange. Communication technology can include simple tools such as meetings, conversations, written documents, whiteboards, and bulletin boards. They can also include more advanced, electronically enabled technologies such as e-mail, text messaging, instant messaging, social media, wikis, or intranet.

The main result of communication and information management planning is the communications management plan, information management plan, and updates to the project management plan, and other project documents are also

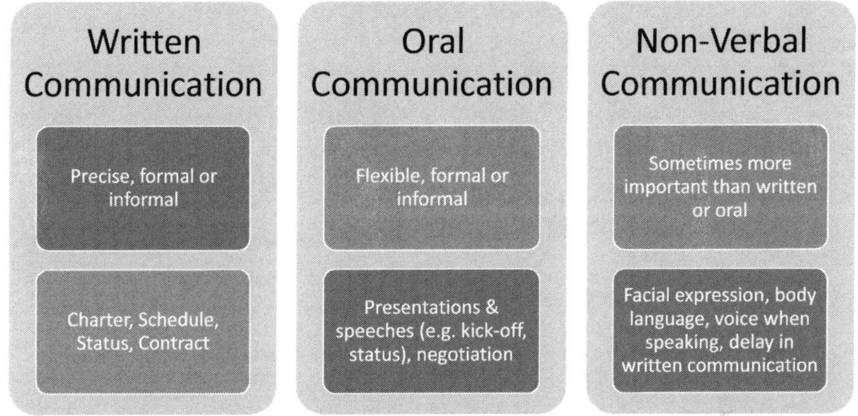

Figure 14.2 Forms of communication.

considered process outputs. Both communications management plan and information management plan are components of the project management plan. As living documents, they should be reviewed and updated regularly throughout the project life cycle as changes occur, especially with stakeholder requirements change.

14.2.1 Creating the Communications Management Plan

Communication often consumes up to 90% of a project manager's activities,[3] which is why it becomes important for project managers to plan, monitor and control communications effectively. The communications management plan describes the overall approach for effective and efficient distribution of relevant information to stakeholders. In the Implementation Phase of a project, the plan strives to distribute the right information, to the right stakeholders, at the right time. Especially on complex projects, the communications management plan attempts to meet the information needs of all stakeholders. The details should be heavily tailored to fit the particular needs of the project and the sponsoring organization.

Within the communications management plan, it is important to outline the following:

- Who will initiate the communication?
- Who will be the recipient?
- What is the key information?
- What medium will be used to distribute the information?

[3] Rajkumar, S., 2010. *Art of Communication in Project Management.* Paper presented at PMI® Research Conference: Defining the Future of Project Management, Washington, DC. Project Management Institute, Newtown Square, PA.

- How frequently will the communication occur?
- What is the desired outcome?

It is also important to determine who is responsible for managing the project information and determining the source of the information. As depicted in Figure 14.3, the RACI tool (Responsible, Accountable, Consulted, Informed) is a useful way to manage the roles and responsibilities of gathering or creating required information.

The communications management plan can include the following components:

- Stakeholder communications requirements
- Information, language, format, content, level of detail, call for action (if any)
- Reasons for distribution of information
- Timeframe, frequency, and responses
- Communication methods and technologies
- People responsible for the collecting of communication
- People responsible for the release of information
- Recipients of information, along with their requirements and expectations of action
- Escalation processes, especially for contentious situations

14.2.2 Developing Information Management Plan

As an important vehicle of change, projects can create a substantial amount of important information and knowledge for the organization. Furthermore, selective information can have long-term applications, for example, for business operational management. Project managers are responsible to develop information management systems and associated processes to effectively and efficiently manage this information. These processes should include the creation of information, capturing and storing of information, providing and managing access to and dissemination of project information, transformation of information into knowledge, refreshing and refining information and knowledge as project progresses, archiving information as appropriate, and on sensitive information, terminating and destroying information based on business needs.

The information management system (IMS) is a set of business processes, technology tools, and policy guidelines to manage the life cycle of information and knowledge, from the initial creation, through use and refinement, to eventual archive or termination.[4] Using IMS, information and knowledge should be managed as a strategic asset and serves a major source of information for a wide variety of project artifacts such as training materials for users, operational instructions for

[4] Williamson, B. & Wu, T., 2019. *The Sensible Guide to Key Terminologies in Project Management*, iExperi Press, Montclair, NJ. Glossary.

						Who is responsible for managing the project info?				
Initiated By	Recipient / Stakeholder	Key Message	Communication Type	Frequency	Desired Outcome	Source of Information	Responsible	Accountable	Consulted	Informed
Senior Leadership Team	Customers	General business updates on launching new capabilities or products	Announcement							
	Company Wide	General business updates on important progress	Announcement or Townhall							
	Board of Directors	Updates on board-level issues	Conference call							
Project Sponsors	Senior Leadership Team		Portal							
	Company Wide		Announcement							
Project Managers	PMO		Announcement							
	Project Team		E-mail							
	Others		Road Shows							

Figure 14.3 Example communications management plan. (Training and consulting content from PMO Advisory LLC. Reprinted with permission.)

back office support staff, information guides for customers, marketing brochures for sales and marketing professionals, and ideas and innovation for generating new concepts and designs for research and development teams.

Insights from the front line. It is important to note that organizations often place more emphasis on the technology aspect of IMS than the processes or policies. For example, organizations may implement sophisticated document management tools with many layers of security and advanced features. This often results in an elaborate system in which few people use. For IMS to be truly valuable, a critical mass of project stakeholders must use the system regularly. Therefore, efforts to encouraging people to adopt the system, such as by conducting regular training and keeping the system simple for common uses, are likely to be far more important.

14.3 Managing and Monitoring Communication and Information Exchanges

During the Implementation Phase of the project life cycle, project managers proactively implement the plans created in the communications management and information management plans. Managing and monitoring communication and information management are processes for managing communication and information to enable the judicious assembly, dissemination and exchange, storage, retrieval, tracking, and disposal of project information. These processes ensure the effective flow of information and use of information in exchanges to create knowledge. The focus is not only on its distribution but also on its retrieval and creation of knowledge. Communication is rarely perfect, and on complex projects, managing communication and information is the major challenge. As such, project managers must remain flexible in their approach to managing the communication of information and sensitive to the changing communication requirements of project stakeholders. Finding the right approach to managing the communication process can be tricky.

There are three different types of communication methods, including push, pull, and interactive methods.

- *Push* methods distribute information directly to specific recipients through mediums such as e-mails, memos, letters, reports, newsletters, blogs, press releases, annual reports, and voice mail.
- *Pull* methods enable recipients to initiate access to information at their own discretion through web portals, intranets, shared databases, wikis, bulletin boards, white boards, and e-learning sites.

- *Interactive* methods emphasize the real-time, multidirectional exchange of information through meetings, presentations, focus groups, phone calls, text messaging, instant messaging, social media, conference calls, and videoconferencing.

Key skills for effective communication include leadership to focus attention and build consensus, political skills to navigate organizational landmines, organizational skills for clear messaging, relationship building, knowledge and information sharing, verbal and written language skills, giving and receiving feedback, and of course, listening. Process outputs include project communications and updates to the project management plan, project documents, and organizational process assets.

14.3.1 Cross Cultural Challenges in Communications and Information Management of Global Projects

Managing communication and information on projects can be complex. However, there are particular factors that make this process all the more challenging for project managers, including teams that are globally dispersed or operate virtually, and managing meetings. Managing a diverse, global team of professionals includes managing language differences and cultural differences. Geert Hofstede[5] outlines cultural differences in terms of five dimensions:

1. Power distance – the ability (or inability) to work with superiors and subordinates
2. Individual versus collectivism – the degree to which people view themselves as an individual or as part of a group
3. Masculinity versus femininity – the degree of preference for alpha male traits such as achievements, heroism, and assertiveness
4. Uncertainty avoidance – the ability (or inability) to work with ambiguity
5. Long-term versus short-term orientation – the connection of the past with future actions and challenges

To meet these challenges, project managers need to understand the cultural norm, speak clearly and use language that is as simple as possible, regularly ask for feedback in order to ensure a proper understanding, and for important discussions, ask listeners to repeat the main points, decisions, or agreements in order to ensure mutual understanding of key information.

With limited time available for project completion, the project manager is also tasked with the challenge of frequent meetings that consume stakeholder time. In order to ensure that meetings are used as a constructive communication tool, rather than a destructive use of time, project managers should always plan ahead

---

[5] Hofstede, G., Neuijen, B., Ohayv, D. D., & Sanders, G., 1990. Measuring organizational cultures: A qualitative and quantitative study across twenty cases. *Administrative Science Quarterly*, 35(2): 286–316.

of meetings by preparing and distributing an agenda with clear meeting objectives. By inviting the most relevant people only and leading the meeting strictly according to meeting objectives and agenda, maintaining meeting start and end times, carefully managing expectations, issues, and conflicts, facilitating problem-solving, and recording meeting minutes, the meeting is more likely to run smoothly and efficiently and result in productive outcomes. After the meeting, the project manager should also distribute meeting minutes with clear action items and owners, preferably within the same business day, in order to provide clarifications, reminders, and a source for follow-up discussions.

14.3.2 Monitoring Communication and Controlling Information Flows

In addition to implementing communication and information management plans, project managers are responsible for the monitoring of communication efficacy and controlling of the efficient and effective flow of communication throughout the duration of the project. These processes strive to enable the information needs of the project, and its stakeholders are met. The ultimate purpose of monitoring communication and controlling information flow is to ensure that the right message is delivered to the right people, in the right format, at the right time, and with the right impact and that it obtains the right response. Selective information on projects can also be sensitive, and project managers should guard their access from unauthorized parties.

When there are deviations from the planned goals as set out by the communications and information management plans and stakeholder engagement plans, project managers must intervene and apply the appropriate corrective actions. This is reflective of the iterative and imperfect nature of communication and information management processes, which often require multiple attempts trial-and-error. Thus, monitoring communication and controlling information flow are part science and part art requiring constant monitoring, analysis, and adapting information according to changes.

Feedback on communication and information management can be collected by means of customer surveys, observations, lessons learned, the issues log, components of project management plan, project performance data and reports, and even tacit sharing of ideas and experiences. When practiced effective, projects will experience few issues, conflicts, changes, and underperformance concerns.

14.4 Transitioning and/or Closing Communication and Information Management

For organizations that manage information as strategic assets, the process of closing communication and information may be the most important. In the Transition and/or Closure Phase, project managers are responsible to ensure relevant information and knowledge are captured, updated, and either archived for future use, transitioning to the next party for continual use, or destroyed the information to prevent downstream risks of unauthorized access.

Depending on the nature of projects, information and communication management closures can be drastically different. For example, for some projects such as event management in which nearly all the major activities end upon the conclusion of the event, project managers should conduct postproject evaluations and archive the relevant information. There are likely important lessons that can be used on future projects. For other projects, such as implementation of a new information technology such as enterprise resource planning or releasing a new product, the real value of information and knowledge created and refined during the project life cycle phases has only started. In these cases, project managers should ensure that key project and project management deliverables are captured and stored in IMS, appropriately updated with the latest project information, coded properly and followed organizational guidelines for easy access by others in the future, transitioned to parties who will be responsible for the ongoing management of the project outcomes and deliverables, and methodically deleted and terminated selective and sensitive information as required by the sponsoring organization.

Insights from the front line. Poor information management can be a source of major risks for organizations, especially in years ahead when there are unintended access and use of information. For example, Company ABC works on client projects, and a legal requirement is the maintenance of a client's data for 3 years. However, this policy of managing information was poorly implementation, and some information remained in the system indefinitely. In a particular legal case, there was a dispute, and the court subpoenaed all the relevant records. The client did not expect that Company ABC still contained its project data, but it did. Even though the information is outdated, it nonetheless negatively impacted the outcome of the case. Company ABC is now liable for the additional loss. From Company ABC's perspective, its lax enforcement of information management policies resulted in reputation loss and financial liability.

14.5 Project Management in Motion

This book is intended for a wide range of audiences and across various industries and functions. Therefore, to balance between competing demands of ease of use, comprehensiveness, modularity, flexibility, and upgradeability, the author has adopted a modular format to this book. This includes the relatively short chapters on key topics. In addition, the appendices include three vital sections:

- Appendix A contains a list of commonly used project management templates for both predictive and adaptive project management approaches.

- Appendix B contains an integrated case study based on a fictitious global company undertaking a number of projects and confronting various project challenges. The case maps closely to the chapters and sections, key concepts, and relevant tools of this book. The goal is to provide instructors, students, and practitioners with a realistic project example to practice and apply the key learnings in the book. For more information and additional case studies in which the author plans to create over time, including potential collaboration opportunities, visit www.optimizepm.com.
- Appendix C contains a glossary of selective terms, reprinting with permission from Dr. Te Wu and Mr. Brian Williamson's book titled "*The Sensible Guide to Key Terminologies in Project Management*", iExperi Press, Montclair, NJ.

Chapter 15

Risk Management – Minimizing Surprises

Summary

Risks abound, especially on large and complex projects. The reason is simple – it is more difficult to plan for "what may" happen especially when risk events are beyond the view of the project team. The detrimental effects of risks are in the news every day. For example, in the Summer of 2017, London just suffered one of the single worst high-rise building fires in its history. The fire at Grenfell Tower killed 72 people and injured 70 more people, and one of the most significant contributors to the rapid spread of the fire is the building cladding. In the aftermath, the U.K. government tested 34 samples of similar cladding from 17 different locations, and 100% of these building materials failed safety tests. To prevent potential disasters, less than 10 days after the fire, the government ordered the evaluation of about 400 buildings impacting nearly 4,000 residents. The very sad part of the story is that flammable building clad responsible for the rapid spread of fire was well known and documented. More recently, in 2018 and 2019, Boeing 737 Max 8 had two crashes about half years apart killing a total of 346 people onboard those plans. In the second case, Ethiopian Airline Flight 302 hurled toward the ground at nearly 600 miles per hour, completely obliterated the plane. In the aftermath, all aviation authorities around the world grounded the plane. Even though blames are contributed by many parties, Boeing's poor project management especially with regard to a number of risky decisions in reducing training, reliance on a single sensor for flight control, and overconfidence in its Maneuvering Characteristics Augmentation System (MCAS) was the prime contributor to those crashes. At the time of writing this book, Boeing's 737 Max 8 remains grounded, and the stock price is nearly 24% below its peak in early 2019.

Even though risks are typically seen as negative in everyday affairs, on projects, risks can also be positive. For example, drug companies often started to mass produce drugs that are poised to be approved by the Food and Drug Administration. Even though the approval has not happened and there is often moderate probability that the approval is not forthcoming, the opportunity to make a profit from another blockbuster drug is too attractive to miss, especially when the patent duration is limited.

This chapter focuses on project risk planning, identifying, analyzing, and responding to project risk. Processes for implementing risk response plans and monitoring and controlling their effectiveness are discussed. The project manager's aim to exploit positive risks (opportunities) and minimize negative risks (threats) is emphasized, including known risks and unknown risks.

This chapter centers its attention on three important risk-related questions:

1. What is so important about managing project risks?
2. How to manage positive risks (opportunities) and negative risks (threats)?
3. How to best analyze risks using quantitative or qualitative tools?

15.1 Importance of Risk Management?

There have been a number of highly consequential failures within the economy and organizations over recent years. Some of the most notable have included the Great Recession that resulted from poor financial management in the financial market, the failure of the Obama Healthcare Website in October 2014, numerous data breaches since 2014 resulting in the exposure of millions of private pieces of information, the Samsung Galaxy Note 7 battery that caught fire and resulted in a loss of at least $17 billion, the Grenfell Tower fire leading to the death of at least 76 and counting, and Boeing's trouble with its 737 Max 8 planes (at the time of writing this book). What these failures demonstrate is that, while a project can be both carefully planned and expertly managed, there is always the possibility of unforeseen events or occurrences that could impact project success.

The goal of the risk management knowledge domain is to minimize surprises, by turning the "unforeseen" into "foreseen". From a risk management perspective, all uncertainties, even positive ones, result in surprises. These surprises reveal the poor control and quality of project management, especially to executives who are impacted by an unexpected outcome. This knowledge domain thus focuses on maximizing value and gains from positive risks (or opportunities) and minimizing the impact and occurrence of negative risks (or threats).

Definition.

1. A *risk* is a potentiality that, if it materializes, can have an impact on one or multiple objectives in a negative or positive manner, in the form of resources, performance, quality, or timeline.[1] This differs from a problem or issue, which is something that has already occurred and is already having an impact.
2. An opportunity is a positive risk, which occurrence is favorable to one or multiple project objectives.
3. A threat is a negative risk, which occurrence can endanger one or more of the project objectives.

To quantitatively compare and prioritize risk, project managers can calculate a risk score, which is determined by the product of the risk's probability of occurrence and magnitude of impact:

Risk score = probability × impact

Risk management is an organized, systematic decision-making process for efficiently planning, assessing, handling, monitoring and controlling, and documenting risk in order to increase the likelihood of achieving project goals and decrease the likelihood that a risk becomes a future problem. It is the process of inquiry about the uncertainty within a project, including questions about the project unknowns (known-unknowns and unknown-unknowns), and results in a common description and understanding of risk. The risk management process assists project managers in identifying existing and potential problems, the resources and strategies necessary to reduce their probability and impact, and the ability to quickly and effectively communicate risk information up and down the management chain. This process maximizes the safety of personnel and provides a structured and systematic review of processes for managing risk, continuous system improvements, and risk communication.

There are four processes for risk management. These include the following:

1. Identifying top-level risks (Ideation and Initiation Phases)
2. Planning risk management (Preparation Phase)
3. Monitoring risk and implementing risk responses (Implementation Phase)
4. Transitioning and/or closing risks (Transition and/or Closure Phase)

Each of these processes will be discussed in the sections that follow.

[1] Williamson, B. & Wu, T., 2019. *The Sensible Guide to Key Terminologies in Project Management*, iExperi Press, Montclair, NJ. Glossary.

15.2 Identifying Top-Level Risks

The identification of risks occurs early in a project life cycle. In the Ideation Phase, risk identification and analysis should be an integral process in the development of ideas, especially at the business case stage. Business professionals should weigh the unknowns and incorporate some measure of them in the financial or performance analysis in the business plan, both the opportunities and threats. At this phase, risks identification tends to focus on broad- and high-level risks, risks that are mainly at the environmental, business, organizational, or product level. As ideas become initiatives and later formal projects, identified risks tend to become more specific and detailed at the project and project management levels.

During the Project Initiation Phase, risk identification and analysis will probably take center stage that coincides with the development of the project charter. At that phase, project managers will work with their core teams to comprehensively and formally identify and evaluate risks in conjunction with scope, schedule, resource, and cost analyses. A good practice is the creation of risk registers, to capture risks and their analysis formally. For larger and complex projects, project managers should also begin the development of a risk management plan.

15.3 Planning Risk Management

In the Preparation Phase, risk management can be an intense set of activities. For sizable projects that involve many stakeholders, project managers should first develop a specialized plan to managing risks on projects. Risk management planning is a process of describing how to perform risk management activities for a project, and the process starts during the Project Preparation Phase. The process is often triggered by additional events that necessitate further planning later on in the project life cycle. To create a risk management plan, project managers work closely with the core team and sponsor and analyze the key project management artifacts including the project charter, project management plan with emphasis on risk-related component plans, internal and external situations, and organizational preferences. Organizational preferences, which include its culture and practices, can disclose its risk appetite or the level of uncertainty a stakeholder or organization is willing to tolerate in anticipation of attaining rewards.[2] By applying sound judgment, conducting data analysis techniques, and facilitating meetings with key stakeholders, project managers can develop a plan for proper risk management on their projects. Another tool to consider is applying a risk breakdown structure (RBS). Similar to a work breakdown structure, an RBS categorizes risks at various levels. With the creation and adoption of a risk management plan, a document

[2] Williamson, B. & Wu, T., 2019. *The Sensible Guide to Key Terminologies in Project Management*, iExperi Press, Montclair, NJ. Glossary.

that provides a baseline of risks and serves as a valuable input to the rigor of risk management, the project can now formally tackle risks in a structured and consistent manner.

Insights from the front line, an example of RBS. An accounting firm is running a global project for building a master data management application. The five levels of associated risk have been outlined using an RBS as follows:

- **Business risk.** Focusing on the functioning of the business as a whole
- **Technical risk.** Focusing on the technology and its interfaces
- **Data risk.** Focusing on data sensitivity and confidentiality issues, such as security and access
- **Management risk.** Focusing on the project and operational management
- **Adoption risk.** Focusing on how the organization and its people accept and use the application

15.3.1 Risk Management Plan

The risk management plan focuses on the relationship between risks and their characteristics, such as risk exposure and project importance to the organization and stakeholder. This enables project stakeholders to be more aware of the threats and opportunities likely to confront the project. The risk management plan forms a component of the project management plan and is an important document, particularly on large, complex, high-stakes projects. The level of detail in this plan varies depending on the project, but it should always contain enough detail for effective planning, intervention, and management of risk.

Components of a risk management plan commonly include the following:

- **Risk strategy.** Overall plan for managing threats and opportunities. This often reflects the organization, the project, or stakeholders' risk culture and appetite and the risk exposure of the project.
- **Approach.** Methods, processes, and tools for identifying, analyzing, and responding to risks. It can also include the extent of planning and analysis, such as trigger point analysis, contingency planning, sensitivity analysis, and so on, to be applied on the project.
- **Roles and responsibilities.** Outline of who will lead, support, and coordinate risk management activities. Key risk owners are also identified and appointed. Risk owners are project stakeholders who are assigned to be the primary person responsible for overseeing and managing the assigned risks.
- **Financial.** Financial funding to buffer the project budget in the event of negative risks becoming a reality. As risks are "probabilities" that may not happen, securing sufficient funding can be difficult.

- **Scheduling.** Coordination of resources at the optimal time for identifying, evaluating, prioritizing, and responding to risks.
- **Tools.** How to capture risks and maintain an active register, what risk category to use, and how to track and record progress. In some organizations, risk audit is an important activity.
- **Special.** How to deal with risks that are not yet known, also called unknown–unknowns.

Even though project managers are often the primary party responsible for risk management, they are not the only party. Some organizations recognize that project managers are rarely the experts in all project components and thus make use of functional experts or dedicated risk management teams. In general, risk management is more robust when there is widespread organizational participation in risk management activities, a thorough risk analysis with expert participation, and shared responsibility by all team members for managing risk. This increased participation also enhances team buy-in to the risk management approach.

15.3.2 Identifying Risks

Studies have found risk identification to impact the transparency of risk directly. Furthermore, the more risk transparency, the better the chances of project success.[3] The process to identify risk starts early on in the project life cycle, where risks are often listed as in the project charter. This is done to ensure that key stakeholders, including the project sponsors and customers, understand the probability of surprises occurring within their projects. In addition to identifying the individual project risks, it is important to understand their characteristics and sources (internal or external, technology or business, portfolio or program or project level, etc.). For larger and complex projects, project managers should consider assigning a designated risk owner, who can be focused specifically on planning and identifying risks.

Good practice. Risks can hold one or a combination of characteristics:

- Risks can exist in any management area, including the ten knowledge domains in the *PMBOK® Guide*.
- Risks can exist in any combination of multiple areas.
- Risks can be positive (opportunities) or negative (threats).
- Risks may or may not occur; they are inherently uncertain.

---

[3] Teller, J., Kock, A., 2013. An empirical investigation on how portfolio risk management influences project portfolio success. *International Journal of Project Management*, 31(6):817–829.

- Risks can be evaluated quantitatively, qualitatively, or a combination of both.
- Risks have no clear boundaries between the various levels of an organization.
- Experienced project managers find the most effective and efficient way to manage risk.

Insights from the front line: From a project management perspective, if risks occur inadvertently and catch the project team by surprise, it always reflects badly on project managers. Even if opportunities were to occur and the project benefits, it is still poor project management because even more benefits can be extracted if the opportunity is planned and exploited for its potential.

At this phase of the project, there are a number of analyses and documents already created or started that can serve as vital inputs to identifying risks. These includes the project charter, project management plan especially on risk-related components, agreements, procurement documentation, organizational risk management assets, and related project management updates. It is especially important for project managers to work with the core team and key stakeholders on risk identification, as their support and eventual ownership of selective risks will become vital for the success of this knowledge domain and the overall project. Furthermore, while this process tends to be most intense during the Preparation Phase, risk identification, analysis, and prioritization should be an ongoing activity throughout the project life cycle. Key outputs from risk identification include creation or update of the risk register, risk report, and updates to other project documents.

15.3.3 Analyzing Risk

Once risks have been identified, it is important that they are prioritized in terms of their likelihood of occurring and their impact on the project. This requires the analysis of risk, which can be performed using both qualitative and quantitative approaches. Each of these approaches to risk analysis will be discussed in turn.

15.3.3.1 Qualitative Risk Analysis

Not all risks are equal, and resources are often limited. As such, risks must be evaluated in terms of their importance to allocate project resources effectively to the highest-priority risks. This process starts with inputs such as project management plan components, project documents, enterprise environmental factors, and organizational process assets. In order to evaluate risks with the information available, tools and techniques such as communication, interpersonal and team skills, data gathering and analysis, expert judgment, meetings, and risk categorization are used. Data analysis techniques often include risk probability and impact assessments, where risks are evaluated according to their likelihood of occurrence and impact of occurrence. In some organizations, additional criteria will be used, such as the team's ability to intervene and manage the risk. The outcome of the risk assessment is a categorization of project risks at low, medium, or high priority as shown in Figure 15.1.

Another common tool in qualitative risk analysis is using expert judgment. This is the inclusion of external experts, usually on complex endeavors in which the internal expertise is limited. Risk categorization techniques, such as RBS, can assist in identifying additional risks and aiding in the general communication of risks. Performing qualitative risk analysis results in updates to a number of project documents.

15.3.3.2 Quantitative Risk Analysis

A quantitative approach to risk analysis is possible on projects where risks are quantifiable or when numerical data are available for evaluation. The process of performing quantitative risk analysis incorporates similar inputs to the qualitative approach, such as components of the project management plan, project documents, enterprise environmental factors, and organizational process assets. Key tools and techniques include expert judgment, data gathering, interpersonal and team skills, representations of uncertainty, and data analysis. Performing quantitative risk analysis will result in updates to various project documents. Examples of quantitative and qualitative risk analysis tools are shown in Figures 15.2–15.4.

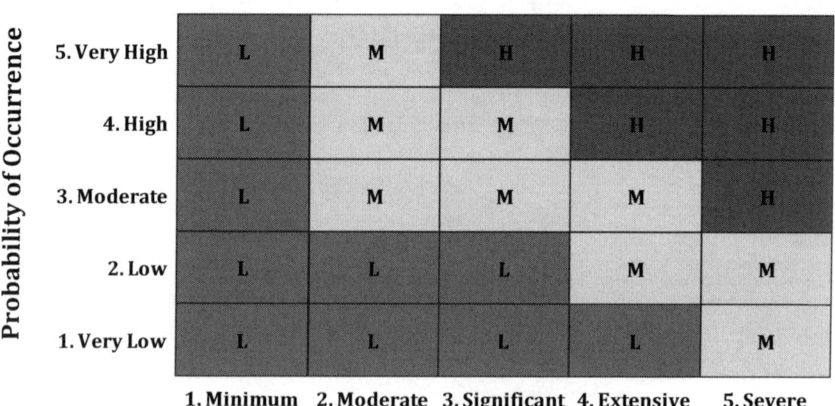

Figure 15.1 Risk analysis.

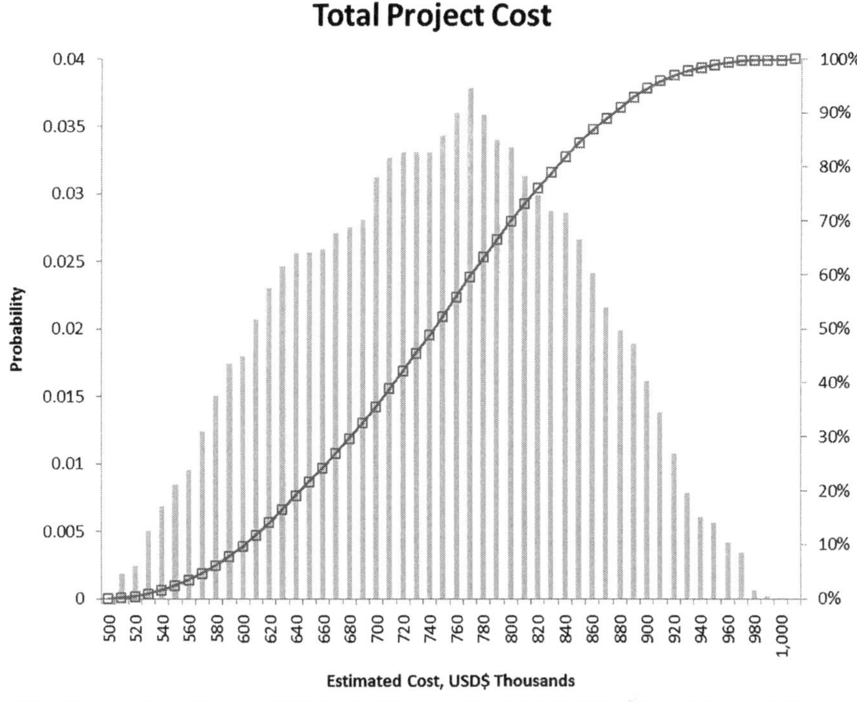

Figure 15.2 Estimating project cost using Monte Carlo simulation. (Training and consulting content from PMO Advisory LLC. Reprinted with permission.)

Figure 15.2 provides an example of a quantitative risk analysis tool. Using a large amount of data to analyze various risks, this example shows a histogram from a Monte Carlo simulation of a project cost estimate. The result shows that the likely cost range is from $500k (almost 0% probability) to $1000 million (100% probability). Thus, if the project sponsors ask for the cost at 50% probability, the project manager should report the project cost at $750k.

A common example of qualitative risk analysis is the force field analysis as shown in Figure 15.3. In this analysis, both the positive driving forces and the negative restraining forces are identified and documented. Using arrows to depict the relative force strength, the analysis graphically illustrates the competing forces. Often used as a decision support tool, the force field analysis provides an easy-to-read summary of the competing forces.

The fault tree analysis (FTA) in Figure 15.4 is another qualitative risk analysis tool to enable a top-down, deductive failure analysis. Here, each of the undesired states of a system or activity is analyzed using Boolean logic to drill down to the specific level of detail. This is commonly used to understand the root causes of defects and failures, especially on complex systems.

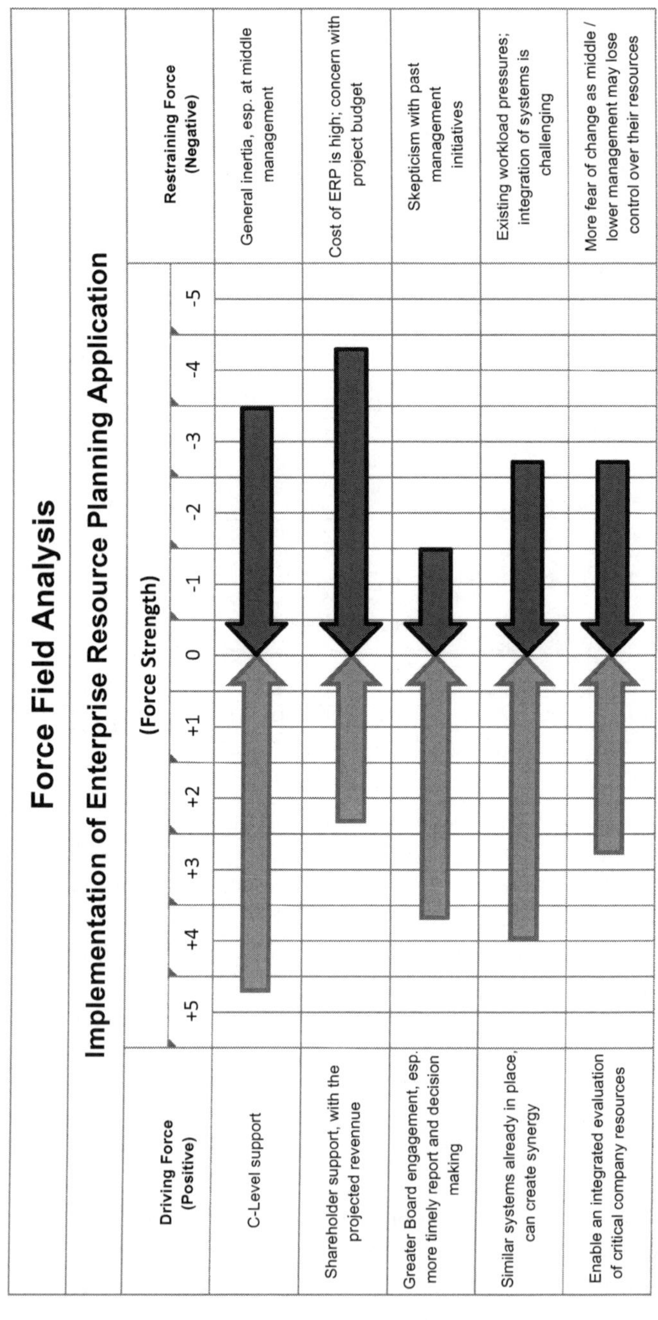

Figure 15.3 Force field analysis. (Training and consulting content from PMO Advisory LLC. Reprinted with permission.)

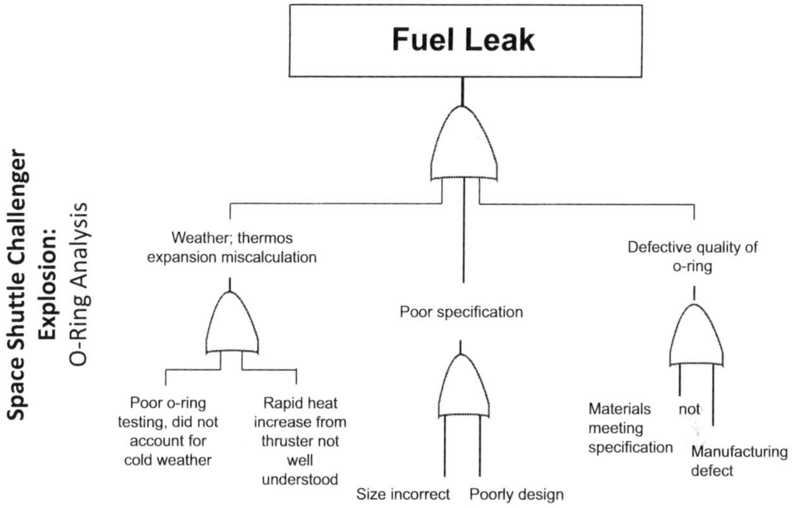

Figure 15.4 Fault tree analysis. (Training and consulting content from PMO Advisory LLC. Reprinted with permission.)

15.3.4 Planning Risk Responses and Contingencies

Even though risks are events that have not yet occurred, project managers should plan appropriate responses and contingencies before they materialize. This way, if a risk does materialize, project managers are ready with appropriate responses. For example, when weather reports predict (with 50% probability) that a severe hurricane will be heading toward New Jersey, the government is unlikely to wait and respond after the hurricane hits. More likely, they will begin planning the response beforehand, evaluating the coastal areas and possibly even evacuating. Organizations such as FEMA, American Red Cross, and even utility companies will probably be involved in the response, in order to ensure sufficient resources are on standby in case of needs. Risks are both positive and negative, and their responses should correspond to their effect (see Figure 15.5).

The goal of the risk response is to identify proper responses in advance, at both the project level and the specific risk level, by identifying options, determining strategies to tackle risks, and deciding on ways to confront risk exposures. This process takes place throughout the project, preparing and documenting the actions and resources necessary to mitigate threats and leverage opportunities should they arise. If and when a risk materializes, the risk owner responsible will lead the execution of the planned response. Responses should always be cost-effective, realistic, and appropriate for the level and complexity of the risk.

To develop risk responses, project managers should work with the core team and utilize the previously created risk analysis and planning documents. These can

THREAT - RISK RESPONSE STRATEGIES	OPPORTUNITY - RISK RESPONSE STRATEGIES
Mitigate: Actions to reduce the probability or impact of a threat	**Enhance:** Actions to increase the probability or impact of an opportunity
Avoid: Eliminate the risk-producing activity entirely by choosing an alternate approach.	**Exploit:** Drive the risk event to occur by changing its probability to 100%, or close to 100%.
Transfer: Take actions to redistribute risk to another area. (This does not relieve the responsibility of tracking and closing the risk)	**Share:** Share risk ownership with an individual or group to influence the opportunity occurrence.
Accept: Accept the risk as stated with no other action.	**Accept:** Accept the risk as stated with no other action.

Figure 15.5 Risk response strategies. (Training and consulting content from PMO Advisory LLC. Reprinted with permission.)

include project charter, project management plan components related to risk management, organizational risk policies, and updated project documents. Given the large variance and complexity of project risks, there are a number of tools and techniques for planning appropriate risk responses.

Contingency planning is a particular type of risk response. A risk contingency plan is completed prior to the risk event but is executed after the risk event occurs. A *risk trigger* usually initiates the execution of the contingency plan. This is a prior event that warns that the risk event will soon occur. Contingency plans are usually put in place when there is no way to further reduce a high risk within a reasonable cost, the risk has a very high probability of occurring, or the risk has a very high impact but a low probability of occurring. Outputs for this process include change requests and updates to the project management plan and project documents.

15.4 Monitoring Risk and Implementing Risk Response

During the Implementation Phase, project management attention shifts toward monitoring risks, especially the high-priority risks. Once risks have occurred, they become events that need to be dealt with as best as possible. Threats become negative events, also labeled as issues, obstacles, problems, or challenges. Opportunities become positive events. In order to maximize the positive benefits of opportunities, or to minimize the negative impact of threats, project managers should implement the risk response as soon as the risk event occurs. If risk response plans are robust, then project managers shift from planning and anticipation to action. If the risk response plans are inadequate, then project managers must adjust the response to react in the best way to the situation.

Alternatively, if the risk response plans are nonexistent, then project managers may be forced to respond entirely reactively with the resources available at the time. In evaluating resources, it is important that project managers monitor the management reserves (part of the overall project funding and budget, and generally controlled by upper management) and contingency reserves (included in cost estimates and the cost baseline, and often controlled by project managers).

To effectively monitor risks and implement risk responses, project managers need to enlist the entire project team and especially the risk owners to support this process. By closely monitoring the risks documented in the risk register and leveraging the risk management artifacts, project professionals apply their expert judgment, utilize interpersonal and team skills, and leverage project management information systems. Effective process will result in potential updates to risk register, change requests, and updates to project documents. Change requests are seldom limited to funding. Other common change types include schedule, resources, scope, quality, and sometimes changes in vendors.

Once risks have occurred and the response plans have been implemented, risk owners are responsible for the continual monitoring of the implementation of the risk response plan, following the identified risks, identifying new risks, analyzing and prioritizing them, and assessing the process effectiveness of risk management throughout the project. In addition, project managers and the project team should be on the lookout for additional risks that can occur throughout the project life cycle. As new risks are identified, project managers should repeat the risk management processes. This includes the following:

- Identifying potential risk
- Evaluating and prioritizing the risk (review, analyze, prioritize, assign risk owner)
- Risk response planning (mitigate, avoid, transfer, or assume assessed risks)
- Executing (response strategies determined in response planning)
- Monitoring new and existing risks and the progress of risk responses

15.4.1 Agile Approaches for Controlling Risk

In Agile approaches, project scope usually cannot be determined precisely in advance. Even when scope can be determined, it can change as the project progresses. Thus, some might consider the Agile approach as a risk management approach to managing this continuous imprecision and uncertainty. In order to deal with risks effectively, Agile teams tend to be smaller than traditional project teams. They are often collocated or exist as multiple clusters each with a particular specialty. What makes Agile effective is often the concentrated effort teamwork. Agile implementation occurs in iterations, also known as sprints. By working on a smaller unit of work (e.g., modular), the goal of Agile is to deliver results faster, fail faster, recover faster, and ultimately realize value faster.

15.5 Transitioning and/or Closing Risk

In the final phase of the project, project managers, risk owners, and key project stakeholders who are involved in managing risks should review and update the risk register comprehensively. The primary objective is to determine how to best manage the residual risks after the project ends. There can be many possibilities with risks; the two most common outcomes are as follows:

1. No more action beyond the project as these risks are no longer material. There can be a number of reasons supporting this. For example, project risks have already occurred, and therefore, they are no longer risks. These risks should be tackled and closed throughout the project life cycle. Or project risks did not occur, and they have dissipated over the project's life cycle. Since they are no longer material, these risks should be closed and no more effort is required.
2. Transition risks to the best organizational units as project risks remain impactful either to the next phase of the project, to the operational environment, or to the sponsoring organization. These can be risks that did not occur during the project but remain material, residual risks that remain in force, or even new risks that arise during the project closure.

For risks that require transitioning, project managers and risk owners should work with project sponsors and governance teams to determine the most appropriate organizational unit or people to continue monitoring these risks. After completing the transition of the remaining risks, project managers should summarize their project risk experiences in the lesson learned and the final project closure report to close the risk management processes formally for the projects.

15.6 Project Management in Motion

This book is intended for a wide range of audiences and across various industries and functions. Therefore, to balance between competing demands of ease of use, comprehensiveness, modularity, flexibility, and upgradeability, the author has adopted a modular format to this book. This includes the relatively short chapters on key topics. In addition, the appendices include three vital sections:

- Appendix A contains a list of commonly used project management templates for both predictive and adaptive project management approaches.
- Appendix B contains an integrated case study based on a fictitious global company undertaking a number of projects and confronting various project challenges. The case maps closely to the chapters and sections, key concepts, and relevant tools of this book. The goal is to provide instructors, students, and practitioners with a realistic project example to practice and apply the key learnings in the book. For more information and additional case studies in which the author plans to create over time, including potential collaboration opportunities, visit www.optimizepm.com.
- Appendix C contains a glossary of selective terms, reprinting with permission from Dr. Te Wu and Mr. Brian Williamson's book titled "*The Sensible Guide to Key Terminologies in Project Management*", iExperi Press, Montclair, NJ.

Chapter 16

Quality Management – Designing It Right

Summary

Achieving quality is no accident. Before even starting the project, project managers must understand the rigor and tolerance of the project deliverables, so quality can be planned from the beginning. Quality is sometimes obvious; for example, a widget designed for consumer use will probably have a more relaxed tolerance than a similar widget built for military use. But quality can also be more subtle such as the perfect alignment of trim on a Lexus vehicle.

This chapter focuses on understanding, designing, and planning for the appropriate quality level for project deliverables. These include both tangible products and intangible results. This chapter introduces the difference between "fit for purpose" (utility) and "fit for use" (warranty).

This chapter examines the three important questions in quality management:

1. Why is project quality management important?
2. What is quality in projects?
3. How to ensure achieving quality deliverables?

16.1 Project Quality Management

Quality is a set of distinctive attributes or characteristics of a product, deliverable, or business result associated with two important concepts: utility and warranty. *Utility* is about being fit for purpose or suitable for function. For example, does the software perform as it was designed to do? *Warranty* is about being fit for use and speaks to the way in which a service is delivered. For information technology systems, this often includes four dimensions, namely, *availability* of the system, *capacity* to support its desired performance, *continuity* of business, and *security.* Project quality management is a knowledge domain responsible for the planning, executing, and controlling project quality requirements in accordance with the organization's quality policy to meet stakeholder objectives. There are four processes in the project quality management knowledge domain:

1. Determining quality (Ideation and Initiation Phases)
2. Developing quality management plans (Preparation Phases)
3. Managing and controlling quality (Implementation Phase)
4. Validating quality to transition or close project (Transition and/or Closure Phase)

In project quality management, it is important that the level of requirements corresponds to the quality rigor. The more specific the quality standards, the more detailed the requirements need to be. It is also critical to manage key stakeholder satisfaction with the project and its progress, by empowering project leads and team members to exercise their expertise. Where possible, project managers should evaluate project situations based on facts and data and adopt a process mentality for project management, where solving problems is balanced with a focus on repeatable and consistent processes. Processes may also have specific entry and exit criteria to ensure quality. These are agreed upon in advance and executed deliberately. In an Agile approach, quality is planned at a high level at the beginning of a project and then at a more detailed level with each iteration or sprint.

Pioneers such as W. Edwards Deming, Joseph Juran M., Genichi Taguchi, and Bill Smith have developed a number of theories, models, and frameworks that have shaped the approaches used today for improving the quality of project management and project deliverables, products, or results. In particular, PDCA is a process tool modified by Deming to:

- **P**lan (identify problems)
- **D**o (test or implement solutions)
- **C**heck (evaluate results)
- **A**ct (implement the best solution) for quality improvement

Juran developed the statistical method called total quality management, which consists of coordinated enterprise-wide efforts to deliver high-quality services and products to customers.[1] While Bill Smith developed Six Sigma, a set of process improvement techniques were designed to reduce process variations.[2]

---

[1] Fletcher, C., 1996. Total quality management: a practical guide. *PM Network*, 10(2): 13–17.

[2] Anbari, F. T., 2002. *Six Sigma Method and Its Applications in Project Management.* Paper presented at Project Management Institute Annual Seminars & Symposium, San Antonio, TX. Project Management Institute, Newtown Square, PA.

Good Practices. Six Sigma includes five steps:

1. **Define quality philosophy and objectives.** Describe processes and understand customer needs.
2. **Measure product and process quality.** Baseline current performance and identify data and collection points.
3. **Analyze data.** Collect and review the data and identify root causes.
4. **Improve processes.** Develop possible solutions for the root cause, and select and pilot the solution.
5. **Control for results.** Create procedures and documentation, train workers, and monitor their performance.

16.2 Determining Quality

During the Ideation and Initiation Phases, the sense of quality tends to be intimated related to preagreed standards such as industry standards or regulatory specifications or organization standards. Common, at this early stage, quality is also intricately related to the scope of the project. For example, constructing a building in San Francisco near the San Bernardino Fault Line will likely face much more rigorous construction standards for earthquake considerations than constructing a similar building in nonearthquake zone.

An important responsibility of project managers in these early phases is to work closely with the idea originators, project sponsors, and other experts to determine the quality parameter required for the project. Questions of utility and warranty should be raised and addressed carefully, as these considerations will directly affect the project scope, schedule, resources, and cost. Quality requirements should be incorporated in the project charter.

16.3 Planning Quality Management

Achieving quality is not easy. It starts with careful planning and diligent implementation. In the planning of quality management, project managers work with the core team and experts to identifying quality needs and requirements of the project deliverables from the stakeholders, evaluating quality requirements (utility and warranty), and documenting them for project implementation and compliance. Stakeholder satisfaction should be a mandatory quality requirement. These stakeholders could be internal, such as project team members, employees, executives, and shareholders, or external stakeholders, such as customers, suppliers, regulatory bodies, and the general public. Project managers are responsible for identifying stakeholders, prioritizing them, understanding their requirements, creating processes and standards to meet their requirements, making trade-off decisions, and implementing the processes and checking results.

The planning of quality management process starts with the available project artifacts such as the project management plan, project charter, organizational policies and guidelines pertaining to quality, and other project documents. By applying tools such as utilizing data gathering and analysis, applying quality test and inspection planning, and conducting meetings with experts, project managers can develop the quality management plan. Since ensuring quality can be costly, project managers often analyze the cost of quality to determine the optimal balance between agreed requirements and anticipated quality or performance. Poor quality can be more costly to remedy in latter project phases. Business judgment can be vitally important as the law of diminishing returns means that cost of prevention will eventually offset the cost of correction. Hardly any complex product or service is defect free, but they can still be "quality products" if they meet the specified requirements.

Two primary types of quality costs are *conformance* costs, which are costs incurred to avoid poor quality, and *nonconformance costs,* which are incurred for correcting poor quality. Prevention costs, such as proactive methods and implementing quality systems, and management costs for training, measuring, testing, and auditing are examples of conformance costs. Failure costs incurred due to poor quality, reputational costs when failure negatively impacts brand equity or firm reputation, and legal costs for product safety issues are all examples of nonconformance costs. The goal is to find the optimal balance between utility, warranty, customer satisfaction, and cost of investment.

Upon completing the process of planning quality management, the project team would create plans and analysis such as the quality management plan, agreed quality metrics, and updates to the relevant project documents. The quality management plan should contain the quality standards and/or methods utilized, project quality objectives, quality management roles and responsibilities, project deliverables and processes, quality control and quality activities, quality tools, process improvement procedures, nonconformance escalations, and quality metrics. Quality metrics are qualitative or quantitative measurements specific to a product or project attributes, which enable comparisons between actual and target results. Examples of quality metrics include failure rate (by severity), defects (per number of units), downtime (per day), training results (pass/fail), ease of use, errors (per line of code), documentation quality, customer satisfaction (scores), requirements (% covered in test plan), repair rate, system response (in seconds), and transaction speed.

16.4 Managing and Controlling Quality

During the Project Implementation Phase, project managers focus on managing the implementation of the project deliverables in accordance with the preagreed quality requirements and standard. Here, project managers are constantly working with the project teams and other stakeholders on quality-related activities. For project deliverables with concrete specifications, such as building a bridge, the quality specifications such as length, width, load tolerance, and height are well specified. While not necessarily easy to achieve, these specifications can be objectively measured, assessed, and compared. For project deliverables that are more subjective, such as design work or ease of performing a workflow, the quality of their functional or esthetic specifications is more subjective and difficult to measure.

Good practices. Quality should be deliberate, intentionally planned, and excellently delivered by following these five best practices:

1. **Customer satisfaction.** Customers are the final judges of satisfaction, and so project managers should understand their expectations and manage them carefully. On difficult projects, customer expectations can be unrealistic given the project constraints. For this reason, it is the responsibility of project managers to involve customers throughout the project at every step. Updates, good and bad, should be promptly communicated.
2. **Empower people and team.** This is especially important on larger projects where project managers cannot be everywhere or know everything. Therefore, it is important to empower project teams to perform within the established standard, method, and processes.
3. **Fact-based management.** Quality should be based on objective data and feedback. This way, project managers can understand and manage variations, understand and improve data and measurement, and adjust course where necessary. Whenever possible, project managers should seek objective measures and standards. Where subjectivity is unavoidable, key stakeholders should be engaged to make determinations.
4. **Process management.** To effectively manage quality throughout the project, project managers must understand, control, and improve processes. These are a series of related activities directed toward achieving the desired results and outputs.
5. **Start early but with the end in mind.** Project managers sometimes fail to look beyond project implementation. On many, if not most projects, the actual utilization and value attained are in the ongoing operations after the project has closed. To address this, it is good practice to incorporate quality processes early on in the project (from the Initiation Phase) and to consider how the product will be used once completed.

16.4.1 Controlling Quality

When there are gaps in quality between the actual and agreed, project managers should intervene to correct and control quality by monitoring and recording project results (interim or final) and making adjustments, as necessary, to ensure the project and product deliverables meet the specified quality requirements. This process can include the inspection of inputs, application of tools and techniques, conducting of activities, quality of deliverables, and robustness of quality management processes, such as reporting systems and quality audits.

Change request is another important mechanism to control quality. These can include preventative actions, such as activities and actions to ensure that the future performance of the project work is aligned with the quality management plan, or corrective actions designed to realign the performance of project work with quality standards. Project managers can also use quality audits to ensure that the project execution meets the agreed standard and requirements. Project audits can be informal or formal, as well as internal or external (see Figure 16.1). Quality control is a process that should be performed by all project team members throughout the project.

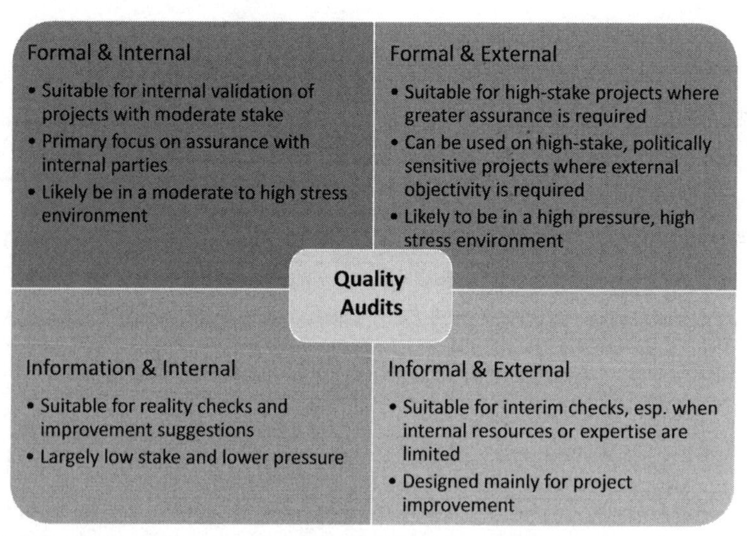

Figure 16.1 Quality audit dimensions.

16.5 Validating Quality to Transitioning and/or Closing Project

In the final phase of the project, project managers should work with the project experts to validate that the quality requirements (e.g., utility and warranty) have been successfully achieved. This is likely a major condition for customers, before they are agreeable to accept and approve the project deliverables. If there remain quality issues, even with customer's acceptance, project managers should transition those considerations to next teams, such as a project team working on the next phase of the project or the operational team, so these quality concerns can be properly managed. Lastly, project managers should capture important lessons learned in the postproject evaluation and the final project closeout report.

16.6 Project Management in Motion

This book is intended for a wide range of audiences and across various industries and functions. Therefore, to balance between competing demands of ease of use, comprehensiveness, modularity, flexibility, and upgradeability, the author has adopted a modular format to this book. This includes the relatively short chapters on key topics. In addition, the appendices include three vital sections:

- Appendix A contains a list of commonly used project management templates for both predictive and adaptive project management approaches.
- Appendix B contains an integrated case study based on a fictitious global company undertaking a number of projects and confronting various project challenges. The case maps closely to the chapters and sections, key concepts, and relevant tools of this book. The goal is to provide instructors, students, and practitioners with a realistic project example to practice and apply the key learnings in the book. For more information and additional case studies in which the author plans to create over time, including potential collaboration opportunities, visit www.optimizepm.com.
- Appendix C contains a glossary of selective terms, reprinting with permission from Dr. Te Wu and Mr. Brian Williamson's book titled "*The Sensible Guide to Key Terminologies in Project Management*", iExperi Press, Montclair, NJ.

Chapter 17

Project Supply Chain Management – Extending beyond the Internal Resources

Summary

In today's competitive environment, few if any organizations perform all the work internally. More likely, organizations focus on what they do best and then find ways to build a supply chain that makes the most sense. Depending on the situation, the supply chain can be composed of business partners who they work with together intimately, suppliers and sellers who manufacture parts or provide services, and outsourcers who essentially take over a part of the organizational activities. This tendency to work with external vendors is likely to be more prevalent with projects. By definition, projects are one-time and often unique endeavors, and organizations are more likely to require outside support as important resources for project implementation. In short, the reality of today's project requires multiple parties, many of which may be external to the organization.

This chapter focuses on planning, managing, and controlling the entire project management supply chain. Although products and services can sometimes be developed internally, these often need to be procured from external entities. Since supply chain management is a vast field, this chapter primarily concentrates on supply chain in the context of projects only and not supply chain management in general. The goal of project supply chain management is to strive toward achieving mutually beneficial relationships and exchanges between all parties.

This chapter examines the three important questions in project supply chain management:

1. How to extend the capability of client organizations through mutually beneficial relationships with external parties?
2. How to align interests between client organizations and sellers?
3. How to manage and control project procurement?

17.1 What Is Supply Chain Management?

Today, most organizations are Lean. This means that they rely on their partners and suppliers for products and services. On projects of any significant size, organizations must work with their supply chain and procure products and services. Project supply chain management is a knowledge domain that addresses how to work with all parties in a continuous chain of activities. The emphasis of project supply chain management is on the procurement of external parties for goods and services and to ensure proper integration with the project team for the successful delivery of projects.[1]

Within supply chain management, the emphasis for projects is the strategic augmentation of project resources beyond the internal capability and capacity. These activities include the acquisition external resources, such as people, materials, or intellectual property, from a multitude of sources. For larger and complex projects, managing supply chain can be a significant responsibility. There are many factors to consider in supply chain management. For example, on smaller and local projects, it is important to consider the timing of resources being utilized on projects and the capability of internal teams and to analyze gaps. The number of suppliers, supplier attributes (such as capability, pricing, and stability), and supplier management processes and guidelines should also be considered. For larger, global projects, there are additional considerations, including global resource acquisition strategy, currency considerations for remunerating vendors, cultural considerations, time zones, and working with virtual or remote teams. Since projects can vary in size, complexity, vendor requirements, governance and regulatory requirements, and physical location, project managers need to tailor their acquisition approach. Advancements toward Lean thinking and Agile methodology for supply chain management, such as migratory models, are yielding new approaches to deal with changing environments.[2]

---

[1] Adopted from the PMBOK Guide (6th ed.) Project Management Institute. 2017. *A Guide to the Project Management Body of Knowledge (PMBOK® Guide) (6th Ed.).* Project Management Institute, Newtown Square, PA.

[2] Potter, A., Towill, D. R., Christopher, M., 2015. Evolution of the migratory supply chain model. *Supply Chain Management: An International Journal*, 20(6):603–612.

There can be a number of challenges to supply chain management, both in sourcing vendors and suppliers and in managing relationships with them during the project. Some of these challenges include identifying how to procure products and services, finding and determining suitable vendors, engaging vendors in Request for Information (RFI), Request for Quotation (RFQ), and Request for Proposal (RFP), short-listing vendors, and vetting vendors to ensure that their capabilities and deliveries match organizational needs. It can also be challenging to negotiate with vendors, to abide by organizational policies for working with vendors, and most of all, to manage the entire supply chain of project activities and capabilities, in turn connecting internal resources with external suppliers.

This chapter on project supply chain management discusses four important processes:

1. Initial evaluation of external resource needs (Ideation and Initiation Phases)
2. Planning project procurement (Preparation Phases)
3. Managing project supply chain (Implementation Phase)
4. Transitioning and/or closing procurement (Transition and/or Closure Phase)

17.2 Initiation Evaluation of External Resource Needs

During the Project Ideation Phase, an important evaluation that arises relatively early in the phase is to determine the resources required to undertake the work required to implement the idea. There are two important dimensions in the evaluation: capability and capacity. Capability refers to the important attributes that the resources need to possess to deliver or support the project activities successfully. For people, capability refers to an individual's competency, skills, knowledge, behavior, and attitude. For physical assets, capability can be an important feature to perform or support project activities sufficiently. Capacity refers to the volume or amount of the resources required to satisfy the project needs. Capability and capacity analysis is an important study, especially on large or mission critical projects in which having sufficient resources is paramount for the projects.

In most scenarios, evaluators tend to look internally first for the quality and quantity of resources required to perform the project, and if the organization has sufficient resources to perform the work, then the evaluation stops. On larger projects, more likely than not, the project requires external resources to augment the internal resources to deliver the project. In some cases, sponsoring organizations may even outsource the entire project implementation to an external party and provide on project management oversight internally.

With the approval of the business cases, the project now enters the Initiation Phase. During this phase, the examination of resources now requires far more detailed examination. In the creation of the project charter, for example, resources are important discussion impacting cost, resources, schedule, risk, and sometimes even scope. Project risks are raised when projects fail to secure important resources. Some resources are scarce and expensive, and significant lead time is required to acquire them. During initiation, projects often onboard core teams, which can serve as an early test to evaluate the organization's readiness to take on the projects. If core resources are missing, and they are difficult to secure, then this can serve as an early warning as the project enters the Preparation Phase of the project life cycle.

17.3 Planning Project Procurement

The planning project procurement is the process of performing a thorough analysis of needs and developing a plan to secure external resources. This process is performed once during Preparation Phase, but depending on the project and project methodology, it can be revisited and updated throughout the project. Specific activities are to:

- analyze and plan for procurement needs (such as capability assessment and resource analysis),
- capture project procurement decisions (such as make versus buy, budget, and funding),
- specify the approach to manage procurements (such as the formal process of request for proposal or a less rigorous process of inquiry),
- work with other relevant teams in the organization (such as supplier management, legal or functional teams, servicing such as subject matter experts),
- identify and contract potential suppliers,
- manage suppliers.

The activity of planning project procurement is more intense in the Preparation Phase, as key decisions will be made. Project managers working with core teams examine the related project artifacts such as project charters, organizational policies and procedures on procuring external products and services, the project management plan, capability and capacity analysis, and other relevant project documents. Applying a combination of techniques such as analysis of requirements, sourcing data, supplier evaluation from previous work, scarcity of resources, and perhaps the most important – sound decision-making, the project team creates the important planning documents such as the procurement strategy, bid documents, Statement of Work (SOW) for selective procurements, source selection criteria, make-or-buy decisions, independent cost estimates, and other procurement management plans and analysis.

For projects with intense procurement needs, the project manager may create the procurement and/or supply chain management plan, which describes how and when to purchase and acquire resources (such as goods, services, and people) from external vendors. Some specific considerations are the type of contract, process rigor or depth, managing risks, supplier selection, and acquisition of resources.

In most cases, project managers will work with additional internal teams, such as supplier management, legal, and probably functional teams for their expertise.

There are four important considerations when planning procurement management. These include the following:

1. Whether to make or buy
2. Contracting process
3. Contract type
4. Transactional purchases versus strategic procurement

17.3.1 Make or Buy Decisions

In determining whether to make a solution in-house or purchase and customize an existing product or solution, project managers should understand their capabilities, environment, and project needs. This includes the time implications for purchasing or developing work, the capability of the project team and organization, the number of suppliers required, roles that internal and external members play, and the availability of sourcing options.

Outsourcing is a form of purchasing in which the customer shifts the majority of the responsibility to the outsourcing company. Companies should consider outsourcing carefully, as there can be many advantages and disadvantages (see Table 17.1). In particular, a strategic advantage is that the outsourcing company can be a strong partner, providing scalability and complimentary capabilities to their customers. Customers can learn how to be more efficient, effective, and innovative from suppliers and vendors. However, there are also many disadvantages. These often include the loss of complementary skills, potential conflict of interest (it is never easy to align the interest of multiple parties completely), and poor management.

17.3.2 Contracting Process

To work with prospective vendors, client companies should clearly state their intentions and motivate vendors to respond. Often the purchasing may be small and organizations already have their chosen vendors. But for larger purchases, especially in government agencies and nonprofits, competitive bidding is a requirement. Three important procurement contracting process and the associated documents often utilized in competitive situations include the RFI, RFQ, and RFP. Depending on the stake of the purchase and organizational policies, organizations can utilize one, two, or all three of these documents and the associated processes:

- The RFI is a request for various pieces of information referring to the products or services. This can also include some indicative pricing, customer references, and seller capabilities. The RFI is utilized when the client company

Table 17.1 Build Internally or Outsource

Make In-House versus Purchase Externally	
Make	*Buy*
• Lower cost of initial production and maintenance • Lack of suitable vendors • Internal capability is strong or at least adequate • Internal resources are available • Core competency area • Desire to have more control over quality • Intellectual property protection, such as proprietary designs • Specific/unique to the organization, difficult to purchase from outside without customization	• Less expensive to purchase • Competitive vendor market • Greater flexibility; will release internal resources to work on other activities • Weak or nonexistent internal capability so making is very risky • Capability not considered important for the organization • Generic products and services requiring little customization • Small quantity required, unprofitable investment of internal time • Insufficient time to make

is seeking to gain a better understanding of the purchase and create a short list of potential vendors. From a vendor's perspective, responding to an RFI is a great way to learn more about the client and position itself as a viable choice.

- The RFQ is often viewed as the next step in the purchasing process. RFQs are used to request more specific price quotations or proposals based on common or standard products and services. The RFQ is generally used to shorten the list of prospective vendors to a manageable few.
- The RFP is the final step in procurement in which the client company works closely with a few most promising vendors to develop specific solutions or products as required. Upon acceptance of the RFP by both parties, the contract, including pricing, often becomes binding.

17.3.3 Contract Type

Client project organizations should also consider the optimal type of contract for their project. Table 17.2 summarizes the most common types and important considerations.

17.3.4 Transactional Purchases versus Strategic Procurement

When procuring products and services, client organizations should recognize the significant differences between a transactional purchase and a strategic procurement.

Table 17.2 Contracting Type

Contract Name	*Description*	*Risk Mainly Absorbed By*	*When to Use from Client Perspective*	*When to Use from Vendor Perspective*
Fixed cost	Buyer and seller agreed to one price and rules/conditions for change order	Seller	Projects are well defined; intervention mechanisms are in place should there be problems with vendors	All cost items are well known; risk can be identified
Time and material (T&M)	Buyer and seller agree to rate, which the seller will bill buyer; material costs are extra	Buyer	Client has strong management capability that can oversee the work; can be less expensive than fixed cost; work is modular and divided across many vendors	High risk projects, often with poorly defined scope and/or subject to change
Cost plus	Similar to T&M, with seller agreeing to a base price plus a profit margin or incentive for seller	Mostly buyer	Scope and cost are well known, few and possibly unique synergy between buyer and seller. There can be multiple cost-plus arrangements where the "plus" can be an additional profit margin, incentive fee, performance award, or even fixed fees. High-risk projects, especially government research projects, often utilize cost-plus contract type	

- *Transactional purchases* are tactical and often involve buying commodity products, services, or other resources. In general, the decision as to which vendor to use is determined by quality and price. If a vendor fails to deliver, then the company can switch to another vendor quickly and without a significant or adverse impact to projects.
- *Strategic procurement* has additional implications, and negative results can significantly impact the project. This is where collaboration and partnering activities are important. In general, strategic procurements occur when there are fewer choices of vendors due to their unique capabilities or products or when the vendors have proven their worth over time. In strategic relationships, vendors should be viewed as partners and can even be considered an extension of the firm. In addition, the business value in strategic relationships is measured not just once but also across the expected life of the relationship.

Even though in nearly all situations, procurement relationship should be managed as partnerships to achieve an optimal value for both sides, strategic procurements are especially important. This is especially in cases when there are few substitutes such as in cases of highly specialized services. Occasionally, suppliers can also shift their strategic importance as major changes impact industries and the macrobusiness environment.

17.4 Managing Project Supply Chain

After preparing for procurements, the next step is to implement during the Implementation Phase. In managing project supply chain, project managers, sometimes working with supplier management professionals in their organization, start the procurement process by determining sellers, ideally from a predevelop list of competing sellers. The process of procuring products and services includes agreeing internally on the vendor evaluation approach and criteria, sharing the procurement scope and objectives to potential sellers, vetting and eventually selecting shortlisting one or more sellers (which may be multiple steps from RFI to RFQ to RFP), obtaining and reviewing vendor responses, conducting research as appropriate, and finally awarding contracts to the sellers.

During this vetting process, project managers should work with their core team, and sometimes the steering committee and sponsor, on the optimal approach for vendor evaluation. Possible approaches include feedback from the vendor's current and previous customers, including internal customers who may have used the vendor before, a comparative review of the seller's products or services, research on vendor reputation for quality, delivery, and postproject services, financial stability analysis (especially important when working with startups), third-party reviews by professional firms, or a site visit to evaluate the seller's capabilities.

For most products and services, there are many potential sellers, and so the client company must follow a number of stages to select the best vendor. The number of stages (such as RFI, RFQ, and RFP) largely depends on four factors, namely:

- Organizational policy (public sectors often have more rigorous processes)
- Internal thresholds (such as cost)
- Products, services, or resources (being procured)
- Specific understanding and knowledge within the organization (RFI is used to collect information)

Despite the implication of these factors, most organizations will follow a process of identifying potential vendors, reviewing their suitability and capabilities, and prioritizing them, before formalizing a contract. To identify potential vendors, project managers should work with procurement experts to analyze sources, including the Internet, other project managers and professionals, and industry resources such as websites, catalogs, trade journals, and professional organizations. Vendors are then reviewed against a list of factors determined during procurement planning. These factors are commonly related to price, scope fit (the RFP usually contains a list of scope items and requirements), quality of response and reputation, on-time performance (here, current and previous customer feedback can be important), ability to work effectively, and seller viability (likelihood of the seller completing the work and supporting it in accordance with the procurement needs).

Potential sellers will not likely volunteer their weaknesses, so it is up to project managers to conduct research and determine areas of concern. Most sophisticated procurement is likely to use evaluation tools and techniques to prioritize vendors. Common techniques include independent estimates, expert judgment, third-party evaluations, and qualitative, quantitative, or hybrid weighing systems.

17.4.1 Contracting

A contract is a legal document that establishes a mutually binding agreement in which the seller is obligated to provide the specified products and services, and the buyer is obligated to pay for them. Anything omitted from the contract is not considered legally binding and puts each party at risk. In complex procurements, there is typically a master contract that lays out the conditions in which the two parties will work together, including payments and disputes. The more specific components of work are covered in the SOW, generally written for a particular product, service, or project component. Other contract components include the scope and deliverables, schedule and milestones, roles and responsibilities, pricing and payment terms, limitations of liability, and pricing with incentives and penalties. The inclusion of incentives and penalties in contracts can improve project time, cost, and performance results.[3]

Table 17.2 lists the three common types of contracts. Each of these also has a number of variations to account for incentives and penalties. For example, in all three types of contracts, there can be incentives for early completion and penalties for late delivery.

Some additional factors to consider when choosing the optimal type of contract include the complexity of scope and requirements, overall risk, urgency and importance of the project, degree of competition available, cost and performance analysis, roles and responsibilities between the buyer and seller, warranty after delivery, extent of subcontracting and its limitations, credit rating of both parties, and project management capabilities.

17.4.2 Managing Vendors

During Project Implementation Phase, the contract is put into practice. The project manager must monitor and control the performance of the agreed-upon work as stated in the contract. Managing vendors is the process of managing the buyer–seller relationships, monitoring the performance of work, making adjustments and changes as appropriate, and ultimately steering the contract toward successful completion. Project managers and core team members leverage the existing project artifacts including project management plan specifically the procurement management plan, procurement documentation such as contracts, project performance, approved change requests, and organizational procurement policies in the management of supply chain. There can be many tools and techniques applied here including performance analysis, inspections and audits, and expert judgment to make informed decisions and advance project progress.

Project managers can also utilize a number of controlling activities on procurements, and they can vary greatly depending on projects and their sponsoring organizations. Some typical activities include the following:

1. Creating performance reports
2. Change control
3. Collaboration
4. Dealing with problems

As a component of the contract, there should be agreement on **performance reports**, including their frequency, type of information, and level of detail. Sellers are responsible for reporting project progress transparently and for raising any

[3] Allen, M., Herring, K., Moody, J., Williams, C., 2015. Project procurement: impact of contract incentives and penalties. *International Journal of Global Business*, 8(2):1.

issues and risks. Project managers on the buyer side are responsible for reviewing these reports and addressing any issues and risks. Changes should be viewed as a normal part of the project, especially those that are large and complex.

Tool: Performance Dashboard. See Template 6: Sample Project Performance Dashboard on page 273 in Appendix for an example of project performance dashboard.

For procurement, **change control** utilization often differs depending on the type of contract. For fixed-price contracts, unless the buyer changes the scope or breaches a clause in the contract, the risk of managing the change generally falls to the seller. Change control is seldom used, unless there are legitimate changes to the contract. For example, in a fixed-cost contract, the seller is experiencing delays due to supply chain disruption. To catch up, the seller project manager decides to hire more expert people to work on the project so work can be expedited. This extra cost is not likely passed to the buyer. The seller project manager should manage the extra resources as an internal project change in the seller's organization.

For time and materials contract types, the cost is more fluid. Change control is still utilized to manage the cost changes, even though it may not negatively affect the contract. **Collaboration** between buyers and sellers is an important ingredient of project success. Both parties should begin early on to align their various interests and develop a partnership mind-set. This way, the contractual arrangement is less about competing gains and more about working cohesively as one project team with a unified set of goals. The two parties should also develop procedures and escalation processes for resolving disagreements before they evolve into more serious conflicts. Effective partnering approaches can be more holistic perspectives that use longer-term outlooks, share power, build trust, and adapt to each other's needs. Here, focusing on issues, rather than positions, ensures mutual commitment to the relationship. Both parties are honest and transparent about shared information and in developing interpersonal relationships.

However, since projects exist under constrained conditions, when multiple parties are involved, some problems may be difficult to avoid. Buyers usually prefer lower prices, but lower prices mean less revenue for the seller. Buyers usually prefer higher quality of products and services, but that quality comes with a price for the sellers who want to maximize profit. These conflicts of interest, whether real or perceived, predispose the buyers and sellers to be suspicious of each other's motives. Project managers should use collaborative and partnering approaches to **deal with problems** and increase the baseline of trust and collaboration.

17.5 Transitioning and/or Closing Procurement

As projects are near their completions, project managers should work closely with their vendor counterparts to review and evaluate project performance thoroughly. Ideally at this late stage in the project, there should be little surprises on their performance, contractual obligation, and process of managing change. However, if there are disputes and disagreements, project managers should raise them sooner so there is time to resolve these differences. This is especially important on strategic procurements since it is likely that the vendor's services may be required in the future.

Once the work is completed, project managers should work with its core team and customers to conduct final reviews and validations of vendor's work. Upon agreement and sign-off, project managers should fulfill the contract obligations such as authorizing the final payments for services and products rendered. In addition to completing the contractual agreements, vendors may continue to play important roles in the sustainment of the project deliverables. In those situations, the project managers should work with the next party, such as operation teams or the next project team, to make proper introductions and transition preagreed activities. During this timeframe, project managers should also complete their evaluation of the vendors and the effectiveness of the parties in the supply chain in their postproject evaluation.

17.6 Project Management in Motion

This book is intended for a wide range of audiences and across various industries and functions. Therefore, to balance between competing demands of ease of use, comprehensiveness, modularity, flexibility, and upgradeability, the author has adopted a modular format to this book. This includes the relatively short chapters on key topics. In addition, the appendices include three vital sections:

- Appendix A contains a list of commonly used project management templates for both predictive and adaptive project management approaches.
- Appendix B contains an integrated case study based on a fictitious global company undertaking a number of projects and confronting various project challenges. The case maps closely to the chapters and sections, key concepts, and relevant tools of this book. The goal is to provide instructors, students, and practitioners with a realistic project example to practice and apply the key learnings in the book. For more information and additional case studies in which the author plans to create over time, including potential collaboration opportunities, visit www.optimizepm.com.
- Appendix C contains a glossary of selective terms, reprinting with permission from Dr. Te Wu and Mr. Brian Williamson's book titled "*The Sensible Guide to Key Terminologies in Project Management*", iExperi Press, Montclair, NJ.

Chapter 18

Leveraging Conflicts – How to Find the Optimal Balance of Conflicts

Summary

Project conflicts are differences that spiraled or elevated beyond mere disagreements, and its impact on project, whether conflicts are constructive or destructive, can be profound. Constructive conflicts are those disagreements that ultimately lead to better solutions. Conflicts can create much needed energy on projects, and when channeled positively, the extra energy can invigorate discussions, create fresh and innovative ideas, push the boundaries of the thought process, and reduce and mitigate group think. More importantly, as project teams overcome conflicts and learned to work with each other, constructive conflicts can be an invaluable and strategic tool to create high-performance teams. For research and development projects or projects that require creative ideas, project managers should intentionally wield the positive power of conflict management.

On the other hand, the negative and destructive power of conflicts is all too real as well. Conflicts stir emotions, and emotions can easily spiral negatively into heated disagreements and confrontations. Destructive conflicts can erode respect, destroy reasoning, obliterate trust, and create a corrosive atmosphere for team work. When unmanaged, destructive conflicts can spiral out of control and pull additional negative energy into the mix that can cause irreparable damage to projects including dooming some projects.

In most of the writing on project conflicts, the tendency is to focus on interpersonal conflicts. While important, there are other major sources of conflicts to consider. This chapter addresses the three important questions in conflict management:

1. Why are conflicts?
2. Why are they important?
3. How to manage conflicts?

18.1 Introduction to Project Conflict Management

In common use, the term conflict has been used in a variety of ways to describe a wide range of disagreements, differences, difficulties, debates, incompatible situations, divergent goals, and struggles. This leads to both an overuse of the term and a general confusion as what are truly conflicts. Here, this book defines conflicts as "escalating disagreements arising from differences in priorities, processes, and personal or organizational views and values. If left alone, conflicts can spiral into larger conflicts with vicious negative cycles feeding and intensifying themselves".[1]

Projects are largely led and performed by people, and where there are people, there is likely to be conflict. This is a phenomenon that impacts all project professionals all over the world. As such, it is critical that project managers have a deep understanding of project conflicts, what they are, why they are important, how they affect the project life cycle and project success, and how they impact the role of the project manager.

Projects today fail at an unacceptable rate. According to the Standish Group,[2] 30% of all software projects are canceled, almost 50% are over budget, 60% are considered a failure, and nine out of every ten are over schedule. However, the effective and efficient management of conflicts could actually have a significant impact on project success.

Project conflicts and their resolutions have been found to have a high correlation with success. Conflicts have been found to deteriorate emotions and the general project atmosphere, reduce productivity, and sometimes even kill the project. When conflicts are well managed, the benefits include reduced groupthink, greater team collaboration and discussion, and more sound solutions and outcomes.

Conflicts can be organized and categorized differently within different frameworks, and one of the most popular topologies was proposed in the 1970s by Thamhain and Wilemon as shown in Table 18.1.

---

[1] Williamson, B. & Wu, T., 2019. *The Sensible Guide to Key Terminologies in Project Management*, iExperi Press, Montclair, NJ. Glossary.

[2] The Economist. November 14, 2004. *Managing Complexity*, 71–73.

Table 18.1 Type of Project Conflicts by Thamhain and Wilemon

#	*Conflict Type*	*Description*
1	Conflict over project priorities	These are conflicts that occur over the sequence of activities and tasks, and it can occur at multiple levels – within project teams, between project teams, and with other groups.
2	Conflict over administrative procedures	These are conflicts over how the project is to be managed. This includes reporting relationships, roles and responsibilities, execution plan, and procedures for administrative support.
3	Conflict over technical options and performance trade-offs	In projects where technology is a consideration, conflicts may arise over technical issues, option analysis, performance specification, and trade-off decisions.
4	Conflict over manpower resources	These are staffing conflicts that can occur, especially on matrix organizations.
5	Conflict over cost	These are conflicts involving estimation of overall or parts of projects, allocation of budget to different parties, and willingness of different parties to share the cost.
6	Conflict over schedule	These conflicts involve timing, sequencing, and scheduling of project-related tasks.
7	Personality conflict	These are interpersonal conflicts that develop around personal differences rather on "technical" issues. These are typically emotion-based and when unresolved can spiral into major firestorms.

Source: Thamhain, H. J., Wilemon, D. L., Spring 1975. Conflict management in project life cycles, *Sloan Management Review*, 16(3), 31.

Thamhain and Wilemon was also able to examine the magnitude of these conflicts across project life cycle, which will be discussed in the next section.

A more recent study conducted by Jehn (1997) presents a similar model of conflicts. In a recent study across six work teams, Jehn categorized conflicts into three types – task, relationship, and procedural or process – and found the following (Table 18.2).

Task conflicts occur 50% of the time, and it is the most common type of conflicts. This is followed by relationship or interpersonal conflict at 30%. The last of the three conflict is procedural or process conflict at 20% of the time.

Table 18.2 Simplified Types of Conflicts by Jehn

Conflict Type	*# of Instances*	*Description*	*Mapping to Thamhain & Wilemon Topology*
Task	71 or 50%	Task conflicts are genuine disagreements over priority of work. For example, in an enterprise resource planning implementation, the representatives from Human Resources, Finance, and Marketing have different priorities for their respective departments. Their inability to reach agreement on the key functions is causing delay for the entire project.	• Conflict over project priorities • Conflict over technical options and performance trade-offs • Conflict over manpower resources • Conflict over cost • Conflict over schedule
Relationship or interpersonal	42 or 30%	Interpersonal conflicts are disagreements between people or groups of people. For example, because of previous negative experience of working with each other. John refuses to be in the same room as Jane. Since both are experts in their respective fields and their collaboration is important, their interpersonal conflict can jeopardize finding an optimal solution.	• Personality conflict
Procedural or process	28 or 20%	Process conflicts are disagreements on "how" to perform work. For example, in a project with parties from multiple companies, they disagree on how to gather project performance data and the content of performance report. When multiple reports were presented to the governance team, the governance team duly asked the multiple parties to present one unified project performance report.	• Conflict over administrative procedures

Source: Jehn, K. A., 1997. A qualitative analysis of conflict types and dimensions in organizational groups. *Administrative Science Quarterly*, 42(3), 530–557.

For the remainder of this section, we will refer to both of these topologies to illustrate how conflicts impact projects across the life cycle. This chapter on project conflict management presents four important processes:

1. Initial evaluation of project conflicts (Ideation and Initiation Phases)
2. Planning to tackle project conflicts (Preparation Phases)
3. Managing project conflicts (Implementation Phase)
4. Evaluating results of conflict resolution (Transition and/or Closure Phase)

18.2 Initial Evaluation of Project Conflicts

Early in project's life cycle, such as Ideation and Initiation Phases, the disagreements tend to be task and process based. According to Thamhain and Wilemon, for example, the most frequent conflict during project formation is conflicts over project priorities. This conflict spikes at the early formation of the project, increases during the project preparation phase, then decreases during the main implementation of the project, and eventually falls below average at closure. The second most frequent conflict spikes during the formation phase, which is an equivalent of the Ideation and Initiation Phases. It should be noted that over the project life cycle, schedule is always more than the average frequency throughout the entire project life cycle, and it becomes the dominant conflict during the main implementation and at closure when time is running out.

During the initial evaluation of project conflicts, project managers and sponsors start by identifying sources and kind of conflicts across the approved projects. Conflicts on certain kind of projects are to be expected. For example, when two or more organizations work on the same project in an interindustry collaboration or in cases of mergers and acquisitions, different companies are likely to have different methods and approaches of implementing projects. Knowing the type of potential conflicts early can help project managers to evaluate the nature and intensity of conflicts. By leveraging the available project artifacts and also working with key stakeholders, project managers apply analytical, interpersonal, and creative skills and judgements to validate their initial assessment and start to prioritize and develop concrete plans to tackle project conflicts.

18.3 Planning to Tackle Project Conflicts

In the Preparation Phase of projects, project managers use inputs such as the project charter, project management plans, and other project updates to start specific plans on how to best tackle project conflicts. The goal is to build on the conflicts that were identified in the earlier Ideation and Initiation Phases as well as newly identified conflicts in the Preparation Phases and develop robust plans to manage these conflicts.

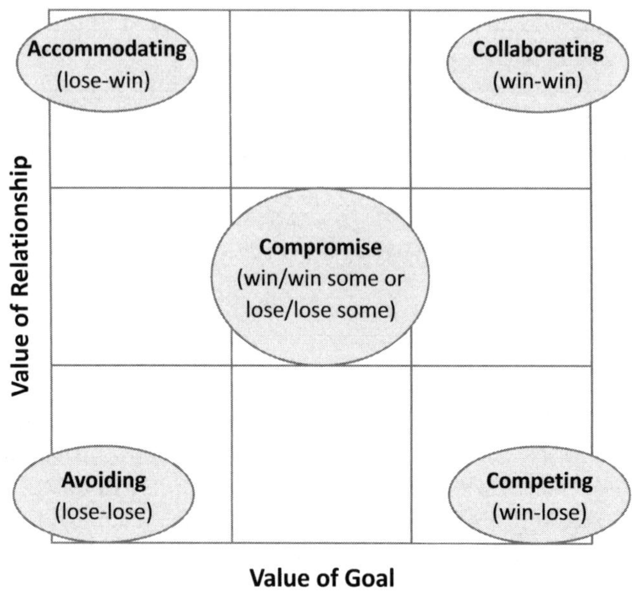

Figure 18.1 Conflict resolution approaches. (Training and consulting content from PMO Advisory LLC. Reprinted with permission.)

A popular method of managing conflicts is derived from Blake and Mouton's Managerial Grid[3] as shown in Figure 18.1. The matrix has two dimensions: value of goal on the x-axis and value of relationship on the y-axis. The value of goal emphasizes on the importance of attaining the individual goals, and when the value of goals is high, individuals are more likely to compete hard to secure their interests. The value of relationship emphasizes the desire to maintain a good affiliation with the other party, and when that desire is strong, especially when the individual goal value is low, one party is willing to put aside their interests and accommodate the needs of the other party.

In the matrix based on these two dimensions, there are five primary approaches to resolve conflicts. When dealing with multiple conflicts, project managers can adopt various styles to deal with different conflicts. Table 18.3 describes the five important styles of resolving conflicts.

[3] Blake, R. & Mouton, J., 1964. *The Managerial Grid: The Key to Leadership Excellence.* Gulf Publishing Co, Houston.

Table 18.3 Conflict Resolution Styles

Conflict Resolution Approaches and Outcome	*Example*
Competing: I **win**, you lose	This style is placing your goal above all else, and the need to win is strong. This should be used sparingly as overuse can significantly damage relationships. For example, there is a product feature that marketing must have, and without the feature, the new product update no longer makes sense. Thus, the marketing department is willing to push very hard to achieve their objective.
Accommodating: I **lose**, you win	This style is placing the relationship with others above achieving your own goal. This may be because your goal is not important. When used in conjunction with competing, relationship can stay in balance, as one party gives into the other party on things that truly matter more to the other party. For example, the marketing team is willing to relent on a low-priority feature in order to smooth hurt feelings on another feature in which they competed vigorously.
Avoiding: Both parties **lose**; can also be neutral	When used conscientiously, avoiding can be an effective strategy. But far more often, avoiding occurs when people become ignorant of conflicts, and thus, there is not a willingness to develop better solutions. For an example of a positive avoidance, the project manager is overwhelmed with work and consciously avoided raising certain issues because there are higher-priority tasks. On the flip side, another project manager avoided issues either of laziness or they seek to minimize disagreements. Yet the conflict does not go away, grows with each heated exchange, and eventually spirals out of control.
Compromising: Both parties win some and lose some	Compromising balances between wins and losses. Both parties did not realize everything they wished, but they were able to secure some wins. More importantly, compromises overcome the current impasse and move the project along. For example, marketing department decided to compromise with operations, so the projects can make progress. Both marketing and operations are not ecstatic about the outcome, but both can live with the results.

(*Continued*)

Table 18.3 (*Continued*) Conflict Resolution Styles

Conflict Resolution Approaches and Outcome	*Example*
Collaborating: I **win** / you win	Collaborating is the ultimate strategy to achieve success. By working together and often not hindered by the existing constraints, both parties are able to find successful solutions to the conflict. For example, marketing and operations department worked together to redefine the problem and value proposition. In the process, both parties were able to achieve what they needed, and both parties win.

18.4 Managing Project Conflicts

Conflicts can be complex phenomena, and they can be composed of both issues and risks, and they are often tracked as issues and risks. On the issue side, they are obstacles that project managers should tackle directly. When these conflict issues are unresolved, they risk elevating from simpler disagreements into destructive conflicts. For example, two area heads started with a simple and genuine disagreement over priorities. But over time and with successful failed attempt at reaching agreements, the conflict's started to be spiraled out of control and how both parties started to clash at an interpersonal level.

It should be noted that in general, there is no "right style", and choosing the right style is situation based requiring a detailed analysis of the situation, including the situation, personalities, and the power dynamics. For managing innovative project, one research by Song, Dyer, and Thieme (2006) found that collaborating (which they called integrating) is the best style, followed by compromising (good), accommodating (fair), avoiding (poor), and competing or forcing (worst).[4]

According to Thamhain and Wilemon, during the main Implementation Phase of projects, schedule conflicts are to be the most frequent followed by technical options, human resources, and project priorities. The conflict over process, cost, and even personality should have been largely resolved as the project team reaches the norming or performing stage of team development.

Throughout the project life cycle and especially during the Implementation Phase, project managers should pay attention to project conflicts throughout the project because conflicts can be sparked by the smallest disagreement. Project are susceptible to a high degree of disagreements because of the multiple constraints.

---

[4] Song, X. M., Dyer, B., Thieme, J. R., 2006. Conflict management and innovation performance: An integrated contingency perspective. *Journal of the Academy of Marketing Science*, 34(3), 341–356.

When unmanaged or poor tackled, projects can be disruptive and endangered. But when managed properly, conflicts can also generate positive energy that drives better problem-solving and create stronger team performance.

18.5 Evaluating Results of Conflict Resolution

During the final phase of the project and as time is running out, schedule conflict is at the top again with all other conflicts subsiding as the project winds down. For project managers, this is a good opportunity to review the performance of conflict management styles and strategies along with issues and risks. Using an integrated and holistic approach to evaluate the conflict management activities critically, project managers should consider conflicts in the postproject evaluation and document the lessons learned before closing the projects.

18.6 Project Management in Motion

This book is intended for a wide range of audiences and across various industries and functions. Therefore, to balance between competing demands of ease of use, comprehensiveness, modularity, flexibility, and upgradeability, the author has adopted a modular format to this book. This includes the relatively short chapters on key topics. In addition, the appendices include three vital sections:

- Appendix A contains a list of commonly used project management templates for both predictive and adaptive project management approaches.
- Appendix B contains an integrated case study based on a fictitious global company undertaking a number of projects and confronting various project challenges. The case maps closely to the chapters and sections, key concepts, and relevant tools of this book. The goal is to provide instructors, students, and practitioners with a realistic project example to practice and apply the key learnings in the book. For more information and additional case studies in which the author plans to create over time, including potential collaboration opportunities, visit www.optimizepm.com.
- Appendix C contains a glossary of selective terms, reprinting with permission from Dr. Te Wu and Mr. Brian Williamson's book titled "*The Sensible Guide to Key Terminologies in Project Management*", iExperi Press, Montclair, NJ.

Chapter 19

Governance Management – Establishing Decision Framework

Summary

One of the most important but often overlooked aspects of managing large and complex project is governance. Governance is "the alignment of project (and also program and portfolio) goals with the sponsoring organization's strategy through sound decision-making processes on authorization, oversight, resource allocation, and change management". Governance is the guidance that forms the boundaries of management practices by setting the project's goals, directions, limitations, and performance framework. This limit or bound can be critically important because of intricate tasks and trade-off decisions confronting complex projects. As such by limiting and guiding management processes, governance greatly simplifies the execution of project and project management processes. Designing an optimal governance framework can also greatly simplify decision-making and increase the efficiency of management processes.

For projects, there can be a multiple layer of governance. Whereas projects exist within programs and portfolios, projects are restricted by the governance at those levels. At the top most level, all project, program, and portfolio governance are guided by the organization governance. Governance can also be applied at a project component level, such as tasks and smaller teams. At those levels, governance tends to focus on the more microconsiderations of individual behaviors within a smaller team environment.

Creating an effective governance can be invaluable to projects as it simplifies decision-making and greatly enhances the overall management of complex, large, and sometimes challenging projects. This chapter on governance addresses these top three questions:

1. What is project governance?
2. How does it work?
3. How to optimize governance framework?

19.1 Principle of Project Governance

The primary role of project governance is to make key decisions effectively. As there are many types of projects often confronting different and sometimes unique situations, establishing a good set of guidelines that work in most situations can be invaluable. As such, governance is embodied by a set of principles rather than overlaid with complicated processes, even though some processes are essential too.

Figure 19.1 shows eight commonly agreed good governance principles. For project management, a common set of principles include the strive toward transparency, responsibility, accountability, and fairness to enable successful outcomes. Here, one can see why management processes are subordinate to the general principles established by governance.

This chapter on project governance discusses four important processes:

1. Initial development of project governance guidelines (Ideation and Initiation Phases)
2. Planning and building governance structure and processes (Preparation Phases)

Figure 19.1 Good governance practices.

3. Operating governance (Implementation Phase)
4. Authorizing project transition and/or closure (Transition and/or Closure Phase)

19.2 Initial Development of Project Governance Guidelines

For organizations with robust project management processes, the question of governance starts in the Ideation Phase, as ideas are developed into business cases. The project governance is typically responsible for the prioritization and authorization of business cases. At the project level, project governance starts mainly in the Initiation Phase during the development of the project charter.

For smaller to moderate size projects, the project is likely to adopt an existing governance structure and associated processes that are aligned with the organization. But for larger, complex, and high-stake projects, sponsoring organizations may wish to develop custom governance that is more fitting for the uniqueness of the project.

In the initial development of project governance guidelines, project managers work closely with sponsors and upper management to establish the guidelines that will oversee projects. Based on the four key principles of transparency, responsibility, accountability, and fairness discussed in the earlier section, project managers can then establish the specific project management processes within the other 11 project management domains (Table 19.1).

The above attributes when combined with additional guidelines such as responsiveness and inclusiveness create a project culture that encourages a certain

Table 19.1 Project Governance Plan Template

#	*Section*	*Description*
A	Basic information	This section contains the essential information about the project including the project name, project manager, project sponsor, start and end dates, and revision history.
B	Purpose	A brief introductory statement defining the purpose of the Project Governance Plan such as: "The Project Governance Plan describes the process that will be followed to execute the Project's governance activities. Its focus is on goals, structure, roles, and responsibilities, and overall logistics of the Governance Board."

(Continued)

Table 19.1 (*Continued*) Project Governance Plan Template

#	*Section*	*Description*
C	Project objectives	This section describes goals and objectives of project governance on the Project such as to ensure that the Project remains aligned with the organization's strategic goals and that interdependencies are managed effectively. It also discusses the role of risk management in governance activities and states the importance of adherence to key policies, procedures, and standards as applicable.
D	Organization structure	This section describes the structure of the Governance Board.
E	Roles and responsibilities	This section lists the members of the Governance Board and specific responsibilities. It describes specific accountabilities for benefit realization, stakeholder communications, and oversight of the Project and its components.
F	Governance decisions	This section describes the decision-making approach the Board will follow. It states how decisions will be documented and communicated to stakeholders and describes an escalation process to follow if the Board does not feel it is empowered to make certain types of decisions. In organizations with specific thresholds of approval, state them here too.
G	Meeting schedule	This section presents an overview of the frequency of Board meetings and notes meetings may be called as needed. It describes the process to follow for meeting logistics and who can attend various meetings. Also, the Governance Board should also plan for urgent or emergency meetings.
H1	Phase gates identification	This section should identify the important phase gates in which major reviews occur. The Project team should have an ample amount of time to prepare for each of these reviews.
H2	Phase gate review requirements	This section states the requirements for Project phase gate reviews. It describes the purpose of these reviews and the items that are covered during each review.

(*Continued*)

Table 19.1 (*Continued*) Project Governance Plan Template

#	*Section*	*Description*
I	Project performance review requirements	This section describes the process to follow for the Board to review overall Project performance at various times. It discusses the objectives of these reviews.
J	Process to initiate new subprojects	For very large projects only. This section describes the process the Board follows to initiate new subproject activity.
K	Assessment of effectiveness	This section describes how Project governance is assessed for its effectiveness in terms of overall delivery of Project benefits and describes who is responsible for this evaluation.
L	Approvals	This section contains the approval of the Project Governance Plan by the members of the Project Board or Governance Board, and other key stakeholders as required.

Source: PMO Advisory Training Material, Copyright 2020.

approach to project implementation. Governance is often established by senior managers, even though project managers can have a significant role in shaping the ideal governance for the project.

19.3 Planning and Building Governance Structure and Processes

In the Project Preparation Phase, in conjunction with the other planning work, project managers should also dedicate time to establish the project governance required for the successful project implementation. One of the main goals is to develop the Project Governance Plan.

Tool: Project Governance Plan. See Template 12: Project Governance Plan on page 281 in Appendix A for an example of governance plan. This template contains many of the common project governance considerations (Table 19.2).

Table 19.2 Example of Governance Principles

#	*Principle*	*Description*
1	Transparency	Transparency is often synonymous with openness, and it involves creating an atmosphere of candidness in which project team members know how decisions are made. It also enables fact-based decision making based on high quality information that is correct, unambiguous, easily accessible. Generally, transparency is led from the top, as each level of the organization brb achieve transparency for subordinate levels. It can also work sideways or multidirectional on more complex projects with both internal and external stakeholders.
2	Responsibility and Accountability	These two terms are often confused, especially because these concepts tend to be used together. The primary difference is that responsibility can be delegated as in the example of assigning a new champion whose job is to promote the project. But accountability cannot be delegated, and even when it is, "the buck still stops here" with the accountable party.
3	Fairness	This is the principle in which all parties and considerations will be treated equally and without bias. For example, if a manager feels that the governance body is making biased decisions unfavorable for the manger's area, then when time comes for execution, the manager may not support the decision fully by holding back on the resources required for implementation.

Source: These examples are adopted from Wu, T. & Chatzipanos, P., 2018. *Implementing Project Portfolio Management: A Companion Guide to Project Portfolio Management.* Chapter 4 Portfolio Governance by Nick Clemens, Project Management Institute, Newtown Square, PA.

19.4 Operating Governance

Project managers are responsible for the routine operation of the project governance functions, and the governance body themselves are accountable for the key decisions and effectiveness of the project governance. The question on the actual implementation of project governance depends on many factors. The art of management is finding what works at a minimum cost and risk. For project management, it is relatively simple to suggest a thorough process of managing project decisions. But project managers must also balance the cost versus the benefits. When project managers establish too much or too heavy processes, the entire governance may appear to be bureaucratic. But when it is lighter than the requirement, governance may be ineffectual.

The key for project managers is to find the optimal balance of what are the essential decisions and guidance required by the governance but also keeping the process relatively light and adaptive. For example, should project managers design the same three-tier governance structure for a $10 million dollar project versus a $1 million project? The answer is that there is not enough information in this example to establish a credible answer. There are other factors of consideration, such as the potential damage if the project were to deliver poor results. For example, the relative worst-case scenario for the $10 million project is a loss of $100 million revenue. But for the $1 million project that upgrades a mission critical system, every minute of poor calculation can be a loss of $1 million. Within a few hours, the penalty of poor calculations can be in the hundreds of millions. In this case, the smaller project clearly requires more governance attention than the larger project.

Governance team may also have specialized focuses that should be designed into the guidelines. For example, on highly innovative projects, the governance guidance may embrace rapid prototyping and adaptive designs. As such, project managers should design rapid escalation processes, so problems can be raised through the organization quickly.

In addition to making key decisions, governance often also have the accountability of oversight and control – making sure the project continues to be aligned with the strategic objectives throughout the project life cycle. When there are deviations, governance should also direct project managers to make adjustments to projects and to steer the project back to the approved paths. Often, governance board members sit higher level in organizations, and thus they have different visibility of the organization and the business environment. Therefore, governance also has the responsibility of observing the macrointernal and external environment and adjust courses as required throughout the Implementation Phase of the project.

19.5 Authorizing Project Transition and/or Closure

As a project completes the work, project governance is accountable to review and approve the deliverables. When there are doubts or uncertainties, governance can order project audits to validate the work being performed. After signing off on the project deliverables, project managers can then start to transition the project to the next team or to close the project.

Prior to terminating the project, project governance should participate in the postproject evaluation exercises and work with the project manager and provide guidance on updating key project documentations and close out of all project activities. In some organizations, the governance team members also have the responsibility of disseminating good project practices and lessons learned to the rest of the organization.

19.6 Project Management in Motion

This book is intended for a wide range of audiences and across various industries and functions. Therefore, to balance between competing demands of ease of use, comprehensiveness, modularity, flexibility, and upgradeability, the author has adopted a modular format to this book. This includes the relatively short chapters on key topics. In addition, the appendices include three vital sections:

- Appendix A contains a list of commonly used project management templates for both predictive and adaptive project management approaches.
- Appendix B contains an integrated case study based on a fictitious global company undertaking a number of projects and confronting various project challenges. The case maps closely to the chapters and sections, key concepts, and relevant tools of this book. The goal is to provide instructors, students, and practitioners with a realistic project example to practice and apply the key learnings in the book. For more information and additional case studies in which the author plans to create over time, including potential collaboration opportunities, visit www.optimizepm.com.
- Appendix C contains a glossary of selective terms, reprinting with permission from Dr. Te Wu and Mr. Brian Williamson's book titled "*The Sensible Guide to Key Terminologies in Project Management*", iExperi Press, Montclair, NJ.

BEYOND PROJECT MANAGEMENT 4

Chapter 20

Working with People – Building, Mobilizing, Managing, and Leading Project Teams

Summary

Hague Sophia, a Byzantine cathedral in today's Istanbul, is an amazing structure. On completion in 537 A.D., it stood as the largest cathedral in the Christian world for almost a thousand years. What is more spectacular from a project management perspective is that the construction took only 5 years, 10 months, and 4 days to complete with a team of just over 10,000 workers. Consider the Second Avenue subway in New York City, which has taken nearly 100 years just to complete Phase 1 – the ancients may know more about project management than we do. Emperor Justinian chose two architects, but Anthemius of Tralles died within the first year. Isidore of Miletus, a physicist, oversaw the completion. Can you imagine leading a team of 10,000 and completing this amazing church in less than 6 years?

This chapter discusses the processes and challenges of implementing projects through people and teams, including creating the team, conducting kick-offs, working with sponsors and governance teams, managing changes in people, and managing stakeholder expectations.

Table 20.1 Project Leadership versus Project Management

Differences between Leadership and Management	
Project Leadership	*Project Management*
• Focus on creating the project vision, setting project direction, and motivating people toward that direction • Demonstrate empathy and care about people and their roles, not just the work itself; emphasize on the behavioral issues; display the ability to influence	• Focus on planning, organizing, executing, and controlling project activities, resources, schedules, budgets, risks, and other project management areas • Demonstrate technical competencies • Emphasis on getting the job done • Display a deep understanding of organizational processes and guidelines

This chapter addresses these three fundamental questions related to working with people:

1. How to build and lead teams?
2. How to motivate people?
3. How to work with project executives?

20.1 Importance of Project Leadership

As the well-known quote from management guru, Peter Drucker states, "*Management is doing things right; leadership is doing the right things*". While both leadership and management are important in projects, they are not the same set of behaviors. Project leadership is the process of guiding the behavior of project stakeholders, especially project team members, through debate or discussion toward accomplishing the project objectives. This is different from management, in which a team is directed toward a point through the use of a known set of management behaviors.[1] The differences between leadership and management are outlined in Table 20.1.

Projects are usually stressful because they exist in highly constrained environments. For this reason, strong project leadership is fundamental to project success. Leaders must direct the project team with clarity of vision, roles and responsibilities, demonstrative project management and technical competency, and the making

[1] Adopted from the PMBOK Guide (6th ed.) Project Management Institute. (2017). *A Guide to the Project Management Body of Knowledge (PMBOK® Guide) (6th Ed.)*, Project Management Institute, Newtown Square, PA.

of difficult choices, preferably in ways that are consistent with the organizational culture. Effective leaders are also able to resolve conflicts, improve teaming and how people work together, apply motivational tools and techniques to encourage team members, and develop trust with stakeholders and manage their expectations. Leaders also care about people as individuals, and not their just work, without losing a focus on getting the project successfully completed.

Leadership, however, is largely a subjective concept, and project managers lead their teams differently depending on their style of leadership. Characteristics of the leader (or project manager), the team members, organization, and environment can all impact the style or approach to leadership. While there are many approaches to leadership defined in the literature, four specific approaches will be discussed here.

20.1.1 The Trait Approach – "Great People" Theory

The trait approach to leadership assumes that leaders are born, and not made, by focusing on a set of natural characteristics. This is often used to justify the "great people" theory. For example, some believe that great leaders, such as Alexander the Great, Genghis Khan, George Washington, or Mother Teresa, were great because of their innate traits. Research by Nadler and Tushman (1990) found support that six natural characteristics are related to great leadership.[2] According to this research, great leaders exhibit the following traits:

1. Excellent intelligence with sound judgment
2. Superior achievement early on in their careers (often in scholarship and sports)
3. Emotional stability and maturity (high EQ)
4. Highly dependable and persistent with drive and energy
5. Socially adaptive
6. Strong desire for status

20.1.2 Behavioral Approach – What Good Leaders Do

The behavioral leadership approach examines what good leaders actually do, rather than who they are. The Ohio State University (OSU) and the University of Michigan conducted two important sets of studies that underpin this approach. Studies by OSU concluded that leaders demonstrate two types of behavior: *Structural behaviors* that establish defined procedures for team members to perform their work, and *Consideration behaviors,* including friendship, trust, and respect, which nurture good relationships between project leaders and team members. The University of Michigan studies also examined behavioral considerations and concluded with

[2] Nadler, D., & Tushman, M. L., Winter 1990. Beyond the charismatic leader: Leadership and organizational change, *California Management Review* 32, 77–97.

two types of behaviors: *Job-centered behaviors* focus on team member activities and tasks, whereas *Employee-centered behaviors* focus on team members as individuals. The conclusions of both studies are similar, with one dimension focusing on work, and another on people. Research shows that effective leaders demonstrate both dimensions.[3]

20.1.3 Situational Approach – When Context Is Important

This approach assumes that, because the situations that project managers confront are different, their leadership skills require a more unique combination based on leader, follower, and situation. This approach consists of three theories (or models) of situational leadership:

1. The *Life Cycle Theory* suggests that effective leadership behavior changes depending on the maturity level of the followers.[4]
2. The *Path-Goal Theory* of leadership suggests that leaders should make rewards available to project team members who exhibit the desirable behaviors.[5] There are four primary types of desirable behaviors in this theory:
   - *Directive behavior* is simply telling team members what to do and how to do it.
   - *Supportive behavior* is the act of being friendly with team members and using personal influence to achieve goals.
   - *Participative behavior* is to seek and incorporate suggestions from team members.
   - *Achievement behaviors* set stretched goals for team members and express confidence and support in them measuring up to the challenge.
3. Tannenbaum and Schmidt (1957) developed one of the most widely quoted articles on situational leadership and how leaders make decisions.[6] The model establishes a broad spectrum of leadership approaches, ranging from management authority for decision-making to providing team members with the freedom to act and decide.
   - The spectrum starts with the project manager making decisions, announcing them, and expecting team members to follow.

---

[3] Hornstein, H. A., Heilman, M. E., Mone, E., & Tartell, R., Spring 1987. Responding to contingent leadership behavior. *Organizational Dynamics*, 15, 56–65.

[4] Hersey, P., & Blanchard, K. H., 1969. Life cycle theory of leadership. *Training and Development Journal*, 23(2), 26–34.

[5] House, R. J., 1971. A path-goal theory of leader effectiveness. *Administrative Science Quarterly*, 16, 321–338.

[6] Tannenbaum, R. & Schmidt, W. H., May/June 1973. How to choose a leadership pattern. *Harvard Business Review*, 162–180.

- Next on the spectrum, project managers communicate their ideas and decisions, discuss pros and cons, and ask team members if they have questions, and then expect team members to follow.
- The third point in the spectrum is where project managers present tentative decisions that can be changed based on team member input.
- Moving further along the spectrum, project managers might present problems, welcome suggestions, and make decisions based on team suggestions, rather than preconceived solutions.
- And finally, project managers might adopt collaborative techniques such as brainstorming, set ground rules, and then ask the team to make the decisions.

20.1.4 Servant Leadership – Placing Focus on the Followers

In servant leadership, the roles of leaders and followers are reversed. In this approach, leadership is not focused on giving orders. Instead, the role of leaders is to make the followers more successful by satisfying the followers' personal and developmental needs, aspirations, and interests.[7] Servant leaders are notable for characteristics like listening more and talking less in order to better understand people's concerns, thus addressing their team's needs. Servant leaders also use persuasion to convince and influence members into action, rather than using orders or positional authority to mandate actions. Sensitive to their environment, servant leaders understand that context sometimes matters more than content, and thus they are keenly aware of their surroundings. Furthermore, servant leaders are empathetic and can make emotional connections by identifying with the feelings, thoughts, and attitudes of others. This way, leaders can help their teams solve their problems appropriately.

20.2 Building Strong Project Teams

One of the most important roles of project managers is to build a strong and competent project team. After planning for resources and estimating their activities, project managers must shift their focus to acquiring these resources. This includes acquiring the necessary skills and capabilities, onboarding team members, and evaluating the project and organizational cultures to find people with the right attitudes and behaviors, in addition to skills.

Project managers often build the core team first. Core team members tend to stay with the project throughout its life cycle and should be competent, trustworthy people who are committed to staying with the project over a longer period of time. To acquire strong project resources, project managers must develop good

[7] Greenleaf, R. K., 1977. Servant leadership: A journey into the nature of legitimate power and greatness. Mahwah, NJ: Paulist Press.

relationships with functional managers, plan for a combination of experienced and inexperienced resources, and persuade team members to work on the project. Project managers should also consider creating a core team, which is composed of excellent professionals with a variety of skills. After creating core teams, project managers can use a similar approach to acquire additional team members. At this phase, project managers can often delegate some human resource tasks to core team members.

Once the team has been acquired, project managers must work with team members as individuals to help them achieve their goals. Since core team members are key to project success, it is important to start by motivating and encouraging them. Common consideration when motivating team members, especially core team members, is to determine the potential benefits for them and their reasons for joining the project team. Most core team members join the project in order to gain new skills, attain more visibility, prepare for promotion, find new challenges, learn about new areas in the organization, or simply network with cross-functional and/or cross-divisional teams.

Next, the team must be onboarded. Depending on the project size, complexity, and team members, project managers can adopt multiple approaches to onboarding. The goal is to ensure that team members understand the project vision and the key objectives. Some project documents can be useful onboarding tools. These include the project charter, project management plan, project requirements, stakeholder analysis, communication plan, and integrated project schedule.

Another best practice is to conduct a project kick-off meeting. The goal of a kick-off meeting is to communicate the project vision, project planning, and constraints to the project team. Despite being called a "kick-off" meeting, these meetings can actually occur throughout the project life cycle, typically at the start of a new phase, or when many new people join the project team. Kick-off meetings typically discuss the project description and basic assumptions, the project approach and methodology, the project scope, schedule, resources and budget, potential issues and risks, and of course, success criteria.

20.3 Developing High-Performance Project Team

As a project enters the project Implementation Phase, the project core team needs to expand. This begins with onboarding the core team members during chartering and then expanding carefully as the project team prepares to execute. The project charter is shared to ensure that members understand the project at a high level. Project managers should learn core team members' personal motives and develop plans for individual improvement. A sufficiently detailed plan (such as a RACI) should then be developed, and team members should be assigned to specific deliverables activities.

However, finding and acquiring the right people on the project team is only the first step to developing a high-performance project team. With team members start working on the project, the project leads must be able to tap into the talent fully that is available to them. Edward, Kinlaw, and Kinlaw (2000) studied the characteristics of high performance in project teams and summarized their findings by addressing six questions[8]:

1. How do project leaders and members understand team development?
2. In what ways, if any, is team development related to project performance?
3. How do the perceptions of project members and leaders about a project's performance compare to the perceptions of key project stakeholders?
4. What is the role of project managers in team development?
5. What are the characteristics of the best or superior project teams?
6. What are the key team development functions that project managers (and other leaders) typically perform?

In the Edward et al. (2000) study, it was found that team development and project performance are interconnected to such an extent that they become the same. As such, project managers and members often do not identify actions that focus on team development as separate from performing work. They also found that team development and project performance are intimately linked, such that higher team development translates into higher project performance. The findings further indicated that the evaluation of project performance by customers, project teams, and leaders is consistent. Therefore, positive perceptions of one party are consistent with the positive perceptions of another party. There was also a nearly unanimous agreement by all stakeholders that project managers are responsible for determining how well projects develop as teams. There was a further general agreement across projects about the characteristics associated with strong project teams. The key characteristics were identified as team focus (working on project needs versus individual wants), communication, empowerment, competence, interdependence, cohesion, commitment, diversity, structure, and recognition. The key specific functions performed by project leaders in developing the project into a strong team were identified as planning team development, initiating team development, integrating project team development and project performance, modeling team work, building cohesion and commitment, coaching, and storytelling.

Outside of this study, other factors related to high-performing project teams typically include the personal characteristics of the project manager and team, and communication considerations. Project managers who have the drive to achieve and personal commitment to the project, who fully understand and accept their responsibilities, and who are willing to listen and consider alternative views

[8] Edward, J., Kinlaw, C. S., and Kinlaw, D. C., 2000. Developing superior project teams, *Scope Management*. A Conference Paper.

typically create high-performing teams. The team itself also needs to have the right balance of skills, be willing to help each other, be focused on continuous improvement, and have effective time management. Finally, effective team communication is important, including sharing information among teams, discussing important (and contentious) topics openly, focusing on tasks rather than personalities when dealing with conflicts, and striving to overcome obstacles.

20.4 Understanding Project Executives

Project executives establish the direction, secure funding and resources, determine the key project success criteria, resolve major obstacles, and approve project changes. Members of this group typically include the project sponsors who are responsible for the day-to-day direction of the project, steering teams (or governance teams) who are responsible for cross-functional engagement and setting direction, the chief project officer who is responsible for the overall project delivery, and financiers who are responsible for funding projects (or sourcing the funding for the project). Project executives often have some, if not all, of the following roles:

- Oversee the development of the project business case, charter, and other key project management deliverables
- Authorize funding and cash flow, often setting up guidelines and tolerances
- Authorize or reject changes to schedule, cost, or scope
- Spend time with project managers on a regular basis to address project concerns
- Work with the project manager in removing obstacles, mitigating negative risks, and exploiting opportunities
- Report to the organizational leadership on project progress
- Support the project manager organizationally (and sometimes politically)
- Authorize project closure
- Celebrate with project managers and teams

Since the role and support of project executives is so critical to the success of projects, poor project sponsorship is one of the most frequently cited reasons for project failure. Therefore, it is important for project managers to work closely with project sponsors and other executives to ensure continual support. There can, however, be difficult situations and personalities, where executives provide unclear direction, a sponsor is overly engaging, a sponsor or executive is not present, or unrealistic expectations are set. Some project executives have many expectations, and these are sometimes incompatible or inconsistent.

Yet, project managers are held responsible if the deliverable does not meet their expectations. In this situation, when the **direction is unclear**, it is important that the project manager clarifies the effort and solution often and early on. Clearly documenting the scope, schedule, budget, and other project constraints, and obtaining agreements and sign-offs, can be helpful later on when deliverables are evaluated. When project **sponsors become too active** in projects, sometimes making decisions without consulting or informing project managers, it can cause confusion and duplication of efforts. Often the intention of the sponsor is to help, but sometimes it is about power play. Project managers should approach this by meeting with the sponsor individually to address these concerns.

The opposite problem can occur too. Some sponsors can become **completely disengaged** and unfairly delegate key decisions to the project manager. When this situation arises, the project manager should also meet with the sponsor individually, and ask for them to become engaged in the project. If that does not work, then sometimes the project manager will need to request the formal authority to make key decisions without the sponsor's involvement. In other cases, creating a steering committee for larger projects and engaging the team, rather than one individual, can assist.

Unrealistic expectations can become a further obstacle, when project executives expect the project manager (and the project team) to achieve **unrealistic or unachievable outcomes**. Here, the project manager should focus on scope management and requirement prioritization and separate the wants from the needs. Agreeing on a prioritization framework early on in the project planning, and applying it diligently, is important. Documenting risks and working with subject matter experts and other project managers or leads to agree on a set of expectations will help to ensure that the executives receive the same message from the project team.

20.5 Project Management in Motion

This book is intended for a wide range of audiences and across various industries and functions. Therefore, to balance between competing demands of ease of use, comprehensiveness, modularity, flexibility, and upgradeability, the author has adopted a modular format to this book. This includes the relatively short chapters on key topics. In addition, the appendices include three vital sections:

- Appendix A contains a list of commonly used project management templates for both predictive and adaptive project management approaches.
- Appendix B contains an integrated case study based on a fictitious global company undertaking a number of projects and confronting various project challenges. The case maps closely to the chapters and sections, key concepts, and relevant tools of this book. The goal is to provide instructors, students, and practitioners with a realistic project example to practice and apply the

key learnings in the book. For more information and additional case studies in which the author plans to create over time, including potential collaboration opportunities, visit www.optimizepm.com.

- Appendix C contains a glossary of selective terms, reprinting with permission from Dr. Te Wu and Mr. Brian Williamson's book titled "*The Sensible Guide to Key Terminologies in Project Management*", iExperi Press, Montclair, NJ.

Chapter 21

Program, Portfolio, Service Management, and Strategic Business Execution

Summary

This chapter discusses the transition between project completion, where the deliverables are created, and the ongoing management of these deliverables through operations and service management. This chapter also extends the traditional concept of project management, which focuses on the implementation of the project, to what happens after the project is completed. This afterlife is generally where value and benefits are achieved.

For many types projects, such as product development or system implementation, the majority of the organizational investment may come after the project completion. For example, a company may have spent a billion dollar developing a new product. But to commercialize the product successfully, the company may spend billions more over the next 5 years to train the sales force, market to the customers, and address customer problems and questions through technical support. In these scenarios, project completion just represents the completion of one stage of the total effort. For the commercialization and ongoing support to be successful, project managers should focus not only on the project implementation but also on the smooth transition to other units after project's completion.

This chapter focuses on these questions that address these three areas:

1. Why should project managers be concerned about the "day after"?
2. What are some related disciplines to project management?
3. How to examine the success of business execution?

21.1 What Happens on the Day After?

What happens on the day after the project is completed? This is a vitally important but often forgotten question to ask. Projects can be different sizes, take on different forms, and have varying levels of complexity. For some projects, the work largely stops upon the completion of the project. For example, if a project is planned for an important annual event, most of the work is complete after the event has taken place. On other projects, some additional work may be required, albeit less intensive than during the project but with far longer duration. For example, a project to implement an information technology infrastructure managed by a third-party vendor may require some further work to monitor performance and manage the vendor for the life of this infrastructure, but the majority of the work is performed during the project. However, the most common scenario in many industries is that a project is only the beginning of the organization's investment. For many projects, the benefits gained from the project occur *not* in the project implementation, but afterward during operations.

EXAMPLE: REALIZING BENEFITS

EXAMPLE 1: GLOBAL MANUFACTURER

A global manufacturer invests close to $500 million over 7 years to implement an enterprise resource planning software. This software will integrate the company's various operational functions. The company will have achieved little to no benefit during project implementation but will expect to save 1% of operational costs (about $20 billion per year) after its implementation.

EXAMPLE 2: TECHNOLOGY COMPANY

A company who develops high-end technology is secretly building a next-generation product that will leapfrog the current competitor's product. Forecasts show that this technology will achieve significant market share gains. After spending $20 million building the product on time, on budget, and with the right functionalities and quality, the company is ready to release. The company will only realize the competitive benefits once the project is completed and the product is released to the market.

Unfortunately, in many companies and situations, project managers have a poor reputation of "tossing the project deliverables over the fence". While project managers are responsible for implementing projects, they are not responsible for managing them after implementation. Instead, the operational team is responsible for the adoption and ongoing management of project deliverables or outcomes. For projects that have a long afterlife, it is vital for project managers to start the project by planning with the end in mind. The most widely adopted framework today for the ongoing management of services (and products) is ITIL®, which stands for Information Technology Infrastructure Library. While ITIL has its roots in information technology, its application today is wide reaching. At the time of writing this book, ITIL just released the 4.0 edition, but some of the components are still being formalized. Therefore, the next section contains information from both ITIL 3.0 and 4.0. ITIL provides a best practice framework for service management, and in ITIL 3.0, this includes service strategy, service design, service transition, service operations, and continual service improvement.

21.2 What Is Service Management?

Service management is the collection of processes and supporting procedures that are performed by an organization to plan, design, deliver, operate, and control services offered to customers. Service comes in many forms, including, but not limited to, building management, customer service management, and information technology. The approach to service management is often based on a particular framework, such as the framework by Kellogg and Nie[1], which connects service with organizational strategy. Even though ITIL® was created within the information technology field, its robustness has made the framework attractive for many other industries and professions. Walt Disney started adopting ITIL best practices in the mid-2000s to ensure that guests checking into their theme parks and resorts have the perfect experience. The Cincinnati Children's Hospital Medical Center led the adoption of ITIL® for a customer-focused, process-oriented, and cost-effective approach in order to minimize business disruptions.

21.2.1 ITIL® 3.0

ITIL 3.0 contains five core publications with 26 processes.[2]

1. The first of these is ***service strategy***, which focuses on helping the organization to achieve its long-term operational goals and objectives. Important

---

[1] Kellogg, D. L., Nie, W., 31 Dec 1995. A framework for strategic service management. *Journal of Operations Management*, 13(4), 323–337.

[2] ITIL Processes. July 20, 2018. *ITIL Docs*. Retrieved from www.itil-docs.com/itil-processes-functions/ on November 10, 2019.

questions should be asked around organizational objectives, what services that operations should provide, likely challenges, stakeholders (e.g., customers) and their needs, the likely demand for these services, financial considerations for providing the best balance of service and quality, and how operations will best meet the needs of the organization and its customers.

2. In ***service design***, the project and operations teams should examine the responses and analysis from Service Strategy and transform them into a concrete plan for delivering business objectives. Typical considerations include *services and service levels*, such as the quality of the services, *availability* of the service, *capacity planning and management*, including the optimal trade-off between service and cost, the *service catalog*, *security management* and how to mitigate security threats, *continuity management*, and *supplier management*.
3. The goal of ***service transition*** is to enable the smooth delivery of services required by organizations in their operational states. The common ITIL® processes in service transition include (1) transition planning and operations, (2) change management and change evaluation, (3) release planning, (4) deployment management, (5) service asset and configuration management, (6) service validation, and (7) knowledge management. While some of these processes are more focused on information technology, the concepts are elastic and can be applied to products as well.
4. The purpose of ***service operation*** is to provide the most effective and cost-efficient services based on the agreed-upon level of services to customers. This includes five common business processes and four functions.

SERVICE OPERATION PROCESSES & FUNCTIONS

FIVE PROCESSES

1. Event management
2. Access management
3. Request fulfillment
4. Problem management
5. Incident management

FOUR FUNCTIONS

1. Service desk
2. Technical management
3. Application management
4. IT or business operations management

5. Organizations and their needs change, so the purpose of the final ITIL® core publication, ***continual service improvement***, is to maintain alignment and ensure that the operational services continue to meet business needs. There are seven processes discussed here, namely, (1) identify the strategy for improvement, (2) define what will be measured, (3) gather data, (4) process the data, (5) analyze the data, (6) present and use the information, and finally, (7) implement improvements.

Collectively, these five publications and twenty-six processes establish the commonly recognized good practices for managing services in ITIL 3.0.

21.2.2 ITIL® 4.0

With the rapid evolution of technology and the development of modular and more Agile approaches to management, ITIL's 4.0 created an expanded framework to both broaden the applicability of service management and increase the agility and modularity of service management by introducing a new concept called service value system (SVS), which represents how the various service management components and activities can work together to deliver value. The processes and activities in ITIL 3.0 form one of the potential streams within the SVS of ITIL 4.0.

To widen the appeal and application of service management, ITIL 4.0 shifted from the concept of management processes to management practices, which are sets of organizational activities with assigned resources to perform work and achieve objectives effectively. ITIL 4.0 contains 34 management practices grouped into three categories: general management, service management, and technical management as shown in Table 21.1.

The organization responsible for managing and developing the ITIL is AXELOS, which is a joint venture between the United Kingdom's Cabinet Office and Capital plc. In addition to ITIL, AXELOS's portfolio of best management practices includes cyber resilience, project management, Agile project management, enterprise agility, program management, risk management, portfolio management, project offices, and value management.[3]

As this is an emerging area, for more information, please refer to the following resources:

- ITIL 4.0 and its release schedule: www.axelos.com/best-practice-solutions/itil
- For a comparison of the 26 ITIL 3.0 management processes with the 34 ITIL 4.0 management practices: https://yasm.com/wiki/en/index.php/ITIL_4_vs_ITIL_V3

[3] AXELOS Best Practice Portfolio Webpage. Retrieved from www.axelos.com/best-practice-solutions on November 10, 2019.

Table 21.1 ITIL 4.0 Management Processes

General Management Practices	*Service Management Practices*	*Technical Management Practices*
1. Strategy management 2. Portfolio management 3. Architecture management 4. Service financial management 5. Workforce and talent management 6. Continual improvement 7. Measurement and reporting 8. Risk management 9. Information security management 10. Knowledge management 11. Organizational change management 12. Project management 13. Relationship management 14. Supplier management	15. Business analysis 16. Service catalog management 17. Service design 18. Service level management 19. Availability management 20. Capacity and performance management 21. Service continuity management 22. Monitoring and event management 23. Service desk 24. Incident management 25. Service request management 26. Problem management 27. Release management 28. Change control 29. Service validation and testing 30. Service configuration management 31. IT asset management	32. Deployment management 33. Infrastructure and platform management 34. Software development and management

21.3 Beyond Project Closure

So now that the project is completed, and it is assumed it was satisfactorily transferred to operations for continual service management. What's next? What is there beyond project management? As discussed in Chapter 1, Introduction, project management is a vast and growing field. This section will discuss some advanced concepts in project management.

But first, let's take a closer look at project management. "Project management" today is a commonly accepted discipline. When most professionals discuss project management, envision this term with a capital "P". This is because to untrained project professionals, Project is a broad umbrella concept as it deals with all

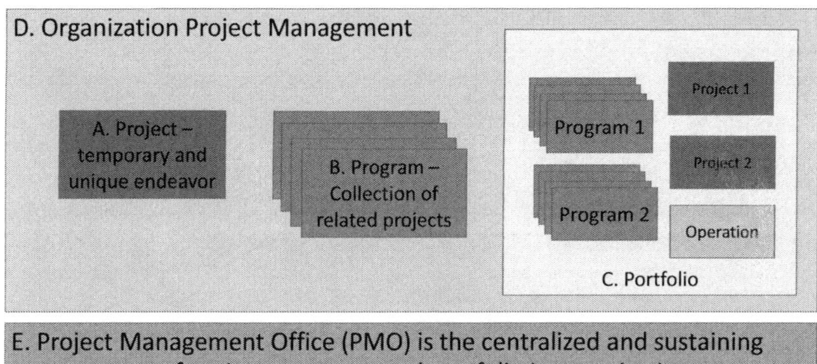

Figure 21.1 Organization project management.

one-time endeavors. But now in the final chapter of the book and as a trained project professional, it is important to examine the project umbrella closer and develop a deeper understanding. Figure 21.1 magnifies the concept of "Project" to illustrate a number of more endeavors within.

As shown in Figure 21.1, there are at least five related endeavors within the world of Projects. These are:

A. Projects with a small cap and the associated project management. *Projects* with a small "p" are generally well-defined endeavors undertaken to achieve a specific objective, such as creating a new product, improving an existing process, or achieving a business goal. The end product is the deliverable or outcome.
B. Program and program management. A *program* is when the endeavor is multifaceted and requires multiple projects to deliver something much bigger. The primary purpose of a program is to create benefits for the organization. For example, an organization that is building a capability to deliver a new kind of service might require a program that consists of five different projects. These projects could be focused on (1) business processes, (2) technology, (3) people, (4) regulations, and (5) marketing. Since organizations often have more ideas than the resources to achieve them, it becomes important for a person or a body of people to have oversight of projects and programs and their contribution to organizational strategic objectives.
C. Portfolio and portfolio management. *Portfolio management* is the bridge between organizational planning and execution, and its primary purpose is to achieve business value.
D. Organization project management. *Organization project management* is then the extension of the project, program, and portfolio capabilities to the entire organization.

E. Project management office (PMO) (can also be program management office and to a lesser extent portfolio management office). The PMO provides the centralized and sustaining management of project, program, and portfolio in organizations.

21.4 Business Execution in Rear View

Why do organizations conduct projects? Now in the final section of the book, we can look at project management in a rear-view mirror. The answer is simple – to achieve business results and optimize the implementation for both effectiveness and efficiency. The general track record shows the dismal track record of project success in many industries. For organizations to achieve greater project success, it's important to remember that projects and their environmental context cannot be separated. In other words, even if an organization perfects its project management, the organization's ability to deliver projects successfully will be hindered by other complimentary areas such as organizational change, process improvement, and performance management to name a few. To achieve sustainable execution excellence, organizations need to make more fundamental changes and adopt a broader framework – a new and emerging discipline called strategic business execution (SBE).

The SBE framework (see Figure 21.2) incorporates aspects of (1) culture, value, and behaviors, (2) enabling competencies, (3) core disciplines, and (4) integrated processes in such a way that execution excellence can be achieved sustainably.

Culture is the first component of this framework, and it is the values, attitudes, and behaviors as the intrinsic software that operates at all levels, including individual, team, organization, industry, and nationally. Execution-oriented values, behaviors, and attitudes are depicted in Table 21.2.

The second framework component is enabling competencies. While there are many enabling competencies for strategic execution, the most important are arguably action oriented, balancing, conflict management, decision-making,

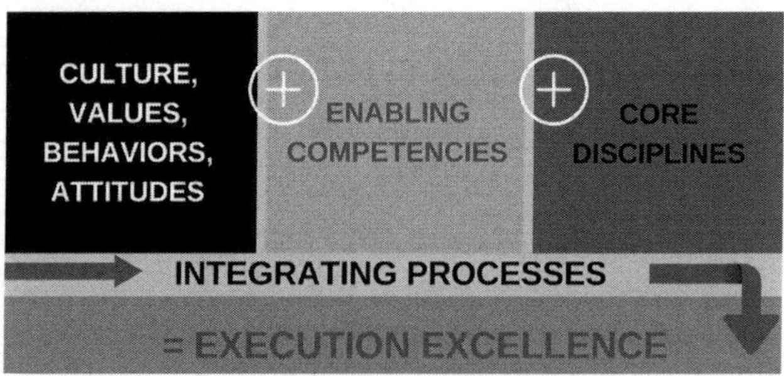

Figure 21.2 Strategic business execution framework.

Table 21.2 Values, Behaviors, and Attitudes of an Execution Culture

Values *Morale Principles and Beliefs*	***Behavior*** *Observable Activities*	***Attitudes*** *Views and Perceptions*
Industrious Resolute Integrity	Respect Teamwork Organization	"Can Do" Optimism Resiliency

delegation, interpersonal savvy, listening, managerial courage, motivating others, and problem-solving.[4]

Organizations also develop their own unique blend of value-added management activities called disciplines, which require regular reinforcement. This is the third framework component. Disciplines are fields of study or practices that codify or institutionalize how activities are performed, both generally within guidelines for good and best practices and specifically related to performance targets with acceptable codes of behavior. Execution-related disciplines include project management, program management, portfolio management, project management office, process improvement, organization development, and execution planning.

Finally, integrating organizational processes serves as the "glue" that connects people, processes, and technologies to achieve common objectives. This is the final framework component for strategic business execution. Some primary integrating processes for execution include communication, knowledge management, dependency management, enterprise issue and risk management, performance management, organization change management, and strategy alignment.

21.5 Project Management in Motion

This book is intended for a wide range of audiences and across various industries and functions. Therefore, to balance between competing demands of ease of use, comprehensiveness, modularity, flexibility, and upgradeability, the author has adopted a modular format to this book. This includes the relatively short chapters on key topics. In addition, the appendices include three vital sections:

- Appendix A contains a list of commonly used project management templates for both predictive and adaptive project management approaches.
- Appendix B contains an integrated case study based on a fictitious global company undertaking a number of projects and confronting various project challenges. The case maps closely to the chapters and sections, key concepts,

[4] Eichinger, R. W, & Lombardo, M. M. for Lominger. 2002. *Career Architect Development Planner*. Minneapolis, MN.

and relevant tools of this book. The goal is to provide instructors, students, and practitioners with a realistic project example to practice and apply the key learnings in the book. For more information and additional case studies in which the author plans to create over time, including potential collaboration opportunities, visit www.optimizepm.com.

- Appendix C contains a glossary of selective terms, reprinting with permission from Dr. Te Wu and Mr. Brian Williamson's book titled *The Sensible Guide to Key Terminologies in Project Management*, iExperi Press, Montclair, NJ.

Appendix A: Selective Project Management Templates

Template 1: Identifying and Analyzing Projects

Introduction

Projects are everywhere. Chances are, you are already a project manager, managing everyday endeavor at work and in personal life. Use this template for a quick exploration of impactful projects in your life by answering the following questions.

Question 1

Have you ever worked on endeavors with the following attributes?

- Temporary
- Fairly unique, either in the outcome or the process of managing work
- Confronting some level of constraints (e.g., time, budget, resources, material resources, etc.)

Come up with up to five memorable projects. Feel free to add rows if you wish to come up with more examples of projects.

#	*Name or Description of Endeavor*	*Category (personal, work, school, etc.)*	*Describe the Desired Outcome or Result*
1			
2			
3			
4			
5			

Question 2

Pick one of the most challenging endeavors from above and provide a broad description of this endeavor that includes the type of work and the constraints (factors that limit the endeavor such as time, effort, resources, material support, leadership, and funding). Also, identify up to five challenges and explain whether the endeavor was successful or *not.*

Project Description
[Insert project description.]

#	*Challenge Name*	*How Did You Tackle Them?*
1		
2		
3		
4		
5		

Question 3

For the endeavor in question #2, if you were to re-do it, what would you do differently? Identify up to three activities that you performed well will do the same and identify up to three activities that you want to improve and would do differently in the future.

A. Activities That Were Done Well

#	*Name*	*Description*
1		
2		
3		

B. Activities That Require Improvement

#	*Name*	*Description*
1		
2		
3		

Template 2: Project Charter

A project charter is a document that represents a formal organizational commitment to execute a project. The project charter is a contract between project manager and project sponsor, in which the sponsor authorizes the project manager to proceed with the continual implementation of the project. Project charter is typically created in the Project Initiation Phase, and its approval signals the completion of Initiation and the start of the Preparation Phase.

Project Name
[Provide the name of the project]

Project Description
[Provide a brief description of the project. Include any compliance/regulatory requirements of this project, for example, that the project be launched by a certain date. If this project is a part of a larger program, please indicate that here. Also if there is a predefined project description, you may provide the reference to that document.]

Project Sponsor	*Project Manager Assigned*	*Core Team Member (if available)*
[Provide the name of the sponsor of this project.]	[Provide the name of the project manager assigned to this project.]	[Provide the name of the key roles required to be a part the core team]

Business Objectives
[Describe in detail what the project aims to accomplish, including qualitative or quantitative metrics whenever possible. Use bullet format.] ...

Business Case Summary
[Concisely describe the business justifications for this project. Use prose format.]

Project Assumptions
[State key assumptions that have been used to inform schedule, scope, and budget estimates up to this point. For example:] • **Scope.** Vendor will provide data input utilities meaning that the development team will not need to develop custom utilities. • **Costs.** System will be built using software already licensed to the company.

Project Guidelines (The content below are examples, please modify to make sure they are appropriate for your project)		
Priority	*Category*	*Guideline*
1	Schedule	**Schedule is the highest priority.** The need to launch and operate the new initiative on or before January 1, 2020, is the highest priority. To meet this deadline, the implementation team should do everything possible including reducing scope, applying more resources, and using manual workarounds.
2	Cost	**Cost limitation.** The company will not spend more than $1.5M in developing and building the new initiative.

(*Continued*)

Project Guidelines (The content below are examples, please modify to make sure they are appropriate for your project)		
Priority	*Category*	*Guideline*
3	Scope	**Minimize customization.** We will "buy" or license viable servicing software and minimize any builds, including customization.
4	Resourcing	**Build core competency.** The company should retain the core competencies of managing and operating the new initiative. This means that we will perform the core activities whenever possible. Where we cannot or should not perform the core implementation activities, it will ensure that mechanisms for critical knowledge transfer will be set up.
5	Performance	**Comprehensive testing.** We must fully test all components of the new initiative (people, processes, and technology) before July 1, 2004. All high severity issues must be resolved before launch.

Project Risks
[State the known risks on this project]

Project Resources
[State the vital and scarce resources required for the implementation of this project. This includes particular skills, technology tools and environment, and other "out of ordinary" needs.]

Project Success: Desired Deliverables or Project Results
[Describe the desired deliverables or outcomes of the project. For each deliverable or outcome, concisely describe the expected quality standard and its priority. For priority, use a scale of **Must** Have, **Should** Have, and **Would** Like To Have. Use bullet format. For example: • **Application form (Must)**. New form must be perceived as easy to use by 90% of the customers involved in the trials. • **Electronic interface between form and back end system (Would)**. Interface must be error free.

Planned Approach
[Describe the high-level approach to complete the project, including steps to jump start the project. It is implicitly understood that situations and changes in the project may change the approach; There is no need to revise the charter with the revised approach.]

Signed and Approved By	
Project Sponsor ______________________	Date: __________
Project Manager ______________________	Date: __________
Program Sponsor ______________________	Date: __________
Program Manager ______________________	Date: __________

Template 3: Project Visioning Tool

Introduction

The high-level design tool (or "project visioning tool" – because you want key stakeholders to "envision" and articulate what the future state will look like) can be used in situations where there is a need to:

- Gain clarity on what the desired end state of the change will be
- Explore alternate paths or options for change
- Agree and build alignment to the desired change
- Understand or convey the urgency and drivers for the change (e.g., create a compelling change story)
- Understand the group or team's strategic vision or make the vision comprehensive and operational

It is a critical starting point for designing the future state – what the organization will look like after the change. Visioning should be approached as an iterative process, and the project manager should continually revisit the vision of the future state to confirm its validity and make updates as necessary.

The outputs of this tool will be dependent on the process and objectives for the visioning but are typically:

- An individual or group representation (whether written or pictorial) of what life or the business environment might be like in the future (vision for future state). The length of time into the future will be dependent on the change being discussed. It could be a few months, a year, or even 10 years.
- An aligned stakeholder group who can each articulate the desired vision and the drivers for this change. Even one stakeholder who does not have a firm

understanding of the future state vision can derail the change – even without meaning to.

- A document summarizing any key decision-making or issues in setting the vision.

When to Use This Tool

Project visioning can be facilitated using a number of different initiatives. The decision regarding which initiative to pursue depends on various factors that include (but are not limited to) the change type, business situation, need for stakeholder alignment, level of agreement on change issue, logistics, and time available. The processes all strive to follow the high-level steps described below, but typically range from:

- Face-to-face interviews with key stakeholders
- A small group workshop (5–10 people)
- A large group workshop or event (15+ people)
- A survey delivered electronically

Instructions

Step 1 – Identify Stakeholders

Identify key stakeholders to be involved in an initiative. This visioning is typically facilitated with key project stakeholders or stakeholder groups that are important to engage and gain support from, such as:

- Initiative owner or sponsor in the business
- A business leadership team
- A project steering committee or wider project team

Step 2 – Describe the Future

A facilitator will set the timeframe for when the future state needs to be achieved. Will it take 6 months, 3 years, or 10 years for the change to take place and the future state to be achieved?

The facilitator asks a series of questions and facilitates discussion between stakeholders to describe the future state – specific to the change that is going to take place. Use the case for change/change drivers as well to focus the future state visioning:

- What do they see? What does work life look like?
- How do people communicate? When do people communicate? How is feedback incorporated?
- How do people work together?

- What sort of technology is available? How do people use it?
- Where do people gather? How do they make decisions? What is the greatest challenge that people have?
- How do people behave? What is right? What is wrong? What is different from the old way?
- How is the future relative to KCG's vision and strategy?

Note that there can be alternate visions of the issue if complex and disparate views preexist.

Step 3 – Capturing the Descriptions of the Future

Stakeholders should record their vision in either written or pictorial form – this may be facilitated discussion where comments, thoughts, and ideas are captured on white boards/flip charts. Facilitators should try and capture an overall vision for the stakeholder group that illustrates the direction for the change. Sometimes a professional illustrator helps turn mental images into drawings that people can extend and modify.

- Benefits of a vision/description of the high-level future state
- Stretches: A good vision challenges the organization to move outside its "comfort zone" to take risks
- Inspires: A well-expressed vision motivates employees by reducing ambiguity and by involving them in leadership's view of the future
- Directs: A vision directs the organization internally and externally through its choosing of the future it wants to create
- Aligns: A vision aligns diverse organizational elements toward a common goal

Step 4 – Deciding on the Final Vision/Description of the High-Level Future State

The facilitator should invite stakeholders to evaluate the pros and cons of the descriptions of the future state using the information that has been developed in the Case for Change to better understand the opportunities and challenges behind different visions (if stakeholders are developing their own vision/description, as opposed to a group developed vision)

Using a method to vote or prioritize the final options, the group should then come to consensus on the single best fit path.

Characteristics of Visioning:

The required characteristics that typically need to exist when finalizing the future state vision ensure that it:

- Takes a long-term view
- Describes the end state in enough detail to take next steps for the detailed future state design
- Involves a collective exercise to aid wider ownership and understanding where possible
- Presents/addresses the initial case for change (e.g., what are the drivers or reasons for the change?)
- Encourages disparate views to be voiced
- Aids decision-making if disparate views exist
- Draws upon deeply held feelings about overall directions to solicit opinions about the future
- Avoids piecemeal and reactionary approaches to addressing problems.

Step 5 – Stating the Vision

State the vision – it could be either a paragraph or a number of bullet points. When completed, the vision should present a democratically derived consensus on what the future state should look like – with key stakeholder input and agreement.

The representation of this consensus:

- Can be used to assess willingness to pay to preserve specific environmental attributes or willingness to accept the loss of these attributes
- Can involve a broad range of participants (in demographic terms)
- Presents the drivers for the change and the foundation elements to communicate to wider impacted stakeholders for the urgency for the change

What makes a change vision compelling?

- **Content.** The right change story, including the value as well as benefits (financial, operational, customer, people)
- **Conviction.** Passion and energy for change
- **Connection.** …with people and with the organization's past and present

PROJECT VISIONING TEMPLATE

In x months or years, our organization will (or our project will deliver…)

1. …
2. …
3. …

Template 4: User Story

A user story is a tool and technique used in Agile methodology to capture a description of a feature from an end-user perspective. A user story describes the type of user, what they want, and why. In some cases, it can also ask the question of why not. A user story helps to create a simplified description of project requirements.

User stories are typically short, simple descriptions of a feature told from the perspective of a user or customer who wants the new capability. The story follows a simple script as shown here:

As a<name or type of user >,

I want<some goal >,

so that<some reason >.

(Optional) otherwise, I would have to <explain alternate process or negative consequence.>

#	*Story Name*	*Description*	*Priority*	*Acceptance Criteria*	*Additional Comments*
1	...				
2	...				
3	...				

Template 5: Product Backlog

A product backlog is a common tool in Agile methodology to document and track the new features or enhancements that customers or users want for a particular project. The goal is to prioritize and execute the currently selected product features, reprioritize, and select the next step of features for the upcoming iterations and so. By constantly identifying new ideas and completing high-priority ones, a product backlog serves as an important tool to manage the dynamic and agile delivery of value-added outcomes to customers.

#	*Task/Feature Name*	*User Story*	*Priority*	*Status*	*Assigned to Iteration #*	*Additional Remarks*
1	...					
2	...					
3	...					

Template 6: Sample Project Performance Dashboard

This template provides an example of project performance reporting. It is shown in a dashboard format, and it ideally captures the latest project performance on a single page. More detailed explanations can be provided for each of the major reporting areas. See notes for an explanation of traditional traffic color health indicators.

Project Name	*Project Manager and Sponsor*	*Reporting Period*
[Provide the name of the project.]	[Provide the name of the project manager and sponsor assigned to this project.]	[Provide the reporting period in which this performance report covers]

Overall Project Health	*Component Project Health*			
	Scope	*Schedule*	*Cost*	*Resources*
[Provide the overall health of the project]	[Determine the health for this component]	[Determine the health for this component]	[Determine the health for this component]	[Determine the health for this component]

Project Summary
[In paragraph form, concisely summarize the current health of the project, and provide justification for the project health reported above.]

Actions/Accomplishments This Reporting Period
[In bullet form, highlight the accomplishments during this period. This can include interim or completed milestones.] ...

Plan for the Next Reporting Period
[Concisely highlight the plan for the coming period, especially include planned completion of milestones] ...

Upcoming Milestones (Date and Health)	*Risks & Issues*
[Provide a summary of the upcoming milestones and its health] ...	[Provide a summary of the ***key*** risks and issues.] ...

Notes: Project health can be reported using many ways. In high structured organizations, reports can be based on earned value metrics. However, to simplify for executive presentations, a common method for health status reporting is using the traffic color codes of red, yellow (or amber), and green. Typically, the colors have the following definitions:

- **Green – Healthy.** There can be many issues and risks, but they are under control by the project manager. No escalation is required.
- **Yellow – Caution.** The project team is working through issues and risks, and there can be potential negative implications. Project sponsors should be warned that their intervention may be required to control the project.
- **Red – Trouble.** Problems or risks have already occurred, and escalation is required. The project manager is working with the project sponsors and its governance to reduce the negative impact. In some cases, an alternate plan will be required.

Depending on the organizations, the traffic colors can be associated with quantifiable calculations of earned value. For example, the subcomponent health of project cost and project schedule can be associated with the Cost Performance Index (CPI) and the Schedule Performance Index (SPI), respectively.

Traffic Color	*Cost Performance Index (CPI)*	*Schedule Performance Index (SPI)*
Green	1.0 or higher	1.0 or higher
Yellow	0.85–0.99	0.90–0.99
Red	Less than 0.85	Less than 0.90

Template 7: Risk Register

Project managers create risk registers to capture and track the most important information of project risks. It is important to note that even though project managers tend to be the de facto owner of risks, they should delegate the responsibility of managing risks to other project stakeholders, especially on large and complex projects.

#	Risk Name	Description	Owner	Status	Priority	Risk Response Plan
1	...					
2	...					
3	...					

Template 8: Issue Log

Projects confront a range of obstacles and challenges that impede project progress. Project managers create issue logs to help them track, monitor, report, and control issues. It is important to note that even though project managers tend to be the de facto owner of issues, they should delegate the responsibility of tackling issues to other project stakeholders.

#	Name	Description	Owner	Status	Severity	Action Plan
1	...					
2	...					
3	...					

Template 9: Project Assumption Log

Project assumptions are project factors or attributes that are assumed to be true for the sake of planning. Without the ability to make assumptions, predictions and planning cannot occur. But the very nature of assumptions is based on estimations and projections, past experiences, or knowledge impart by other project stakeholders. Assumptions are therefore risks, and they can change. When assumptions change, they become issues. Project managers should create assumption logs to help them manage this special category of risk. It should be noted that there is a myriad of assumptions on projects, and with only limited resources, project managers should be selective on choosing the assumptions that they need to track.

#	Name	Description	Owner	Status	Potentiality of Change	Implication of Change	Response Plan
1	...						
2	...						
3	...						

Template 10: Project Change Request Form

Purpose	The primary purpose of this tool is to capture, analyze, and manage changes after achieving agreements with the authorized stakeholders.
Project Phase	Can be used throughout the project life cycle to managing changes after achieving agreement. In predictive approach, changes tend to occur more often after the Preparation Phase.

I. Project Information

Project:		Project ID:	
Project Mgr:		Date:	

Requestor:		Role:	
Email:		Phone Ext:	

II. Change Description

Change Request #:			Affected Product Version #:	
Affected Project Discipline:				
Priority:	☐ High	☐ Medium	☐ Low	
Type:	☐ Enhancement	☐ Requirements Changed	☐ Correction	
Documented Testing/Deployment Plan?				
Description of Change:				
Business Benefit:				

Implications of Not Making the Change:			
Alternatives:			
Est. Budget Impact:		Est. Schedule Impact:	

III. Impact Assessment

Area	*Details of Impact*
Functional Scope	
Schedule	
Effort	
Budget	

IV. Request Status

Status	*Date*	*By Whom*
☐ Opened & Logged		
☐ Returned to originator for clarification		
☐ Assigned		
☐ In Testing		
☐ Completed		

V. Authorization

Accepted:	☐ Yes	☐ No	Authorized By:	
Date:			Signature:	

Template 11: Postproject Evaluation Template

Purpose	The primary purpose of this tool is to serve as an internal examination of the project progress and capture the key lessons learned from the project, including areas of strengths and improvement.
Project Phase	Mainly during Project Transition and/or Closure. For larger projects, this template, with some modifications, can be used at the end of major phases or completion of a subproject.

I. Basic Information

Provide the following basic information about the project including:

Project Name	
Project Manager	
Project Sponsor	
Start to End Dates	

II. Status and Comments Column

Complete the Status and Comments column. Indicate:

- Yes, if the item has been addressed and completed
- No, if the item has not been addressed, or is incomplete
- Partial, if there are open components of this item
- N/A, if the item is not applicable to this project

Provide comments or describe the plan to resolve the item in the last column.

#	*Item*	*Status*	*Comments/Plans to Resolve*
1	Have all the product or service deliverables been accepted by the customer?		
2	Are there contingencies or conditions related to the acceptance? If so, describe in the Comments.		

(*Continued*)

#	*Item*	*Status*	*Comments/Plans to Resolve*
3	Have all the key stakeholder groups been informed of the project completion?		
4	Has the project been evaluated against each performance goal established in the project performance plan?		
5	Has the actual schedule cost of the project been tallied and compared with the approved cost baseline?		
6	Have all the groups required for ongoing operations been identified and communicated of the project closure and their ongoing responsibilities?		
7	Has the management of the ongoing/operational teams formally accepted responsibility for operating and maintaining the product(s) or service(s) delivered by the project?		
8	Has the documentation relating to operation and maintenance of the product(s) or service(s) been delivered to, and accepted by, operations management?		
9	Have training and knowledge transfer of the operations organization been completed?		
10	Has the project documentation been archived or otherwise disposed as described in the project plan?		

III. Dos and Don'ts

Please answer the following three questions:

1. What activities were done well on this project and they should be continued?

2. What activities required improvements?

3. On a scale of 1–3, where 1 is no, 2 is maybe, and 3 is yes, is the project success in accordance with the key stakeholders?		
Stakeholder Name	*Rating (1–3)*	*Comments*

IV. Signatures

The signatures of the people below signify an agreement that the key elements within the Closeout Phase section are complete and the project has been formally closed.

Position/Title	*Name*	*Date*	*Phone #*

Template 12: Project Governance Plan

I. Basic Information

Project Name	
Project Manager	
Project Sponsor	
Actual Start Date	
Approved End Date	
Revision History	

II. Purpose

[A brief introductory statement defining the purpose of the Project Governance Plan such as "The Project Governance Plan describes the process that will be followed to execute the Project's governance activities. Its focus is on goals, structure, roles and responsibilities, and overall logistics of the Governance Board."]

III. Project Objectives

[This section describes goals and objectives of Project governance on the Project such as to ensure that the Project remains aligned with the organization's strategic goals and that interdependencies are managed effectively. It also discusses the role of risk management in governance activities and states the importance of adherence to key policies, procedures, and standards as applicable.]

IV. Organizational Structure

[This section describes the structure of the Governance Board.]

V. Governance Board Roles and Responsibilities

[This section lists the members of the Governance Board and specific responsibilities. It describes specific accountabilities for benefit realization, stakeholder communications, and oversight of the Project and its components.]

VI. Governance Decisions

[This section describes the decision-making approach the Board will follow. It states how decisions will be documented and communicated to stakeholders and describes an escalation process to follow if the Board does not feel it is empowered to make certain types of decisions. In organizations with specific thresholds of approval, state them here too.]

VII. Governance Board Meeting Schedule

[This section presents an overview of the frequency of Board meetings and notes meetings may be called as needed. It describes the process to follow for meeting logistics and who can attend various meetings. Also, the Governance Board should also plan for urgent or emergency meetings.]

VIII-a. Phase Gates Identification

[This section should identify the important phase gates in which major reviews occur. The Project team should have an ample amount of time to prepare for each of these reviews.]

VIII-b. Phase Gate Review Requirements

[This section states the requirements for Project phase gate reviews. It describes the purpose of these reviews and the items that are covered during each review.]

IX. Project Performance Review Requirements

[This section describes the process to follow for the Board to review overall Project performance at various times. It discusses the objectives of these reviews.]

X. Process to Initiate New Subprojects

[For very large projects only. This section describes the process the Board follows to both initiate new subproject activity.]

XI. Assessment of Effectiveness

[This section describes how Project governance is assessed for its effectiveness in terms of overall delivery of Project benefits and describes who is responsible for this evaluation.]

XII. Approvals

[This section contains the approval of the Project Governance Plan by the members of the Project Board or Governance Board, and other key stakeholders as required.]

Appendix B: Case Study – Whole World Enterprise (WWE)

Purpose

To complement the book, the author created this integrated case study that incorporates concepts from all the chapters of the book. The goal is to provide instructors and students with a realistic project example to practice and apply the key concepts in the book.

This case study is organized by sections, and each section corresponds to the corresponding chapter in this book. For example, Section VIII. Project Integration Management is mapped to Chapter 4 of this book: Project Integration Management. One of the many goals of the author is to create a series of these case studies across various functions and industries that would enable readers, instructors, and students to apply project management to those respective areas.

There are actually two separate project examples covered by this case. The first case is based on an enterprise resource planning (ERP) project covers Sections I through VII. The second case is a specific project within the larger ERP implementation designed for the remaining sections (VIII–XXI). The basic information presented in Section I applies for both project examples.

For more cases supporting this book, visit www.optimizepm.com.

I. Introduction to Whole World Enterprise

A. Company Overview

Name	Whole World Enterprise (WWE)
Industries served	Retail/wholesale Internet Consumer electronics Edutainment (Education Entertainment) Publishing
Geographic area served	Worldwide, with heavy emphasis in the United States. Current worldwide revenue is as follows: • North and South America: 51% • United States: 70% • Canada: 20% • Mexico: 5% • Rest of Americas (presence in Brazil and Argentina): 5% • Europe: 28% • Asia: 14% • Middle East and Africa: 5% • Australia: 2%
Headquarter	Saratoga Springs, New York
Executives	Mr. Darren (Wai Ming) Zhang, CEO (since inception of firm in 1997) Ms. Adriana Holmes, CPO (joined in September 2012)
Revenue	US $92.1 billion (2019), increase of 12%
Profit	US $860 million (2019), increase of 340%
Employee	170,800 (2019), increase of 10%
Main competitors	Alibaba Group, Amazon.com Inc., Apple Inc., eBay, Facebook Inc., Google Inc., Hewlett-Packard, Hulu, International Business Machines Corporation, Microsoft Corporation (specially Lynda.com), Netflix, New York Times, Spotify, Times Inc., Wal-Mart Stores, and many other Internet and retail companies

B. Business Overview

Founded in 1997 at the dawn of Internet commerce, Whole World Enterprise (WWE) started a virtual eCommerce company striving to dominate various educational product categories. It started with children's products (books, toys, videos, programs, and even digital electronics). It then rapidly moved up the demographic curve including self-help, productivity enhancing tools and equipment, and eventually into a professional office stationery and equipment including computing devices. It also developed a strong institutional market in which WWE sells directly to businesses and institutions on their entire range of products and services. As it achieved a significant market in the basic products, it started shifting, making much of its products including educational programs, documentaries, and featured films, which largely address one or more of the social concerns of our time. In 2010, it also rapidly expanded into the digital space by offering cloud services for the educational market, software, and devices. By 2019, the company had achieved a revenue of over US $92 billion worldwide and was considered one of the fastest growing businesses in the world.

The company is guided by five principles:

1. **Obsession with lifelong learning and "maximizing human potential".** Its motto "reach higher" resonates well with people who aspire to achieve more.
2. **Passion for solving problems.** A vast majority of its products and services are designed to address challenges and to make life more efficient and effective. Its documentaries are full of ideas to solve problems.
3. **Strive to "do good".** While the company is a for-profit organization, social missions such as educating the most vulnerable citizens in all countries that it serves is a strong indication of its commitment to "doing the right things", even at the expensive of profit. Until recently, the company had largely been unprofitable, even though its investors believe in the mission of the company.
4. **Desire for creativity and innovation.** The company focuses on the future and has a long-term outlook. Hence, it encourages its people to think differently and look beyond the near-term.
5. **Fixation on getting things done.** The balance toward a long-term outlook is the preoccupation with achieving results. The firm views its ability to execute its strategies, deliver its projects, and bring value to its shareholders and customers as a vital competitive advantage.

The company serves its retail customers largely through its retail website and focuses on quality products, price, trustworthy reviews (only purchasers who have a complete profile can provide product reviews), and convenience such as fast delivery. In 2014, it also started some retail stores that serve as an educational hub and also serve as a community hangout spot with product demonstrations and quality speakers. For institutional customers, the company developed a strong supply chain management tools to interface with the institutional customer's purchasing and inventory management systems that seamlessly connect WWE as a part of their business customer's value chain. In 2017, the company started professional business training streaming services that sell directly to companies and individuals. The streaming business became so successful that it's often viewed as a platform business.

WWE competes head-on with Amazon.com and Alibaba in multiple product categories. Comparatively, WWE products are more expensive and somewhat slower in shipping speed once Amazon.com introduced Sunday shipping. WWE's product selection is also more limited with about 20 million stock keeping units (SKUs) versus Amazon's 340 million. But WWE views its more limited SKUs as a positive trait because WWE rigorously selects its products, which has created a strong following. WWE understands that most of its customers are confused by product varieties. By providing fewer but higher-quality selections, studies have shown that WWE has a more loyal customer base while retaining a higher price. From a transaction perspective, it is estimated that Amazon.com experiences about 53 orders per second or about 4.5 million transactions per day. For WWE, the volume is significantly less, but it still achieves about 17.5 orders per second or 1.5 million transactions per day. This works especially well with its institutional customers. But the competitive pressure has always kept the company on its edge.

While the first four guiding principles are vastly interesting, this case study primarily focuses on the fifth principle – "getting the right things done".

C. The Burning Platform – The Need for Project Management

In 2011, shortly after the 14-year anniversary of the firm's founding, WWE experienced a near-death experience. Due to the rapid growth of the company, in March 2010, the company decided to integrate the firm by implementing an enterprise resource planning (ERP) tool to break down the barriers across the various areas of the company. The goal was to improve internal efficiency and communication and also to centralize the management of its products, people, and customer accounts. Unfortunately up to that point, the company focused mainly on expansion and not internally on its management capabilities, especially project management. Its internal systems and processes are stitched together by a huge list of patchwork items, and it was breaking at the seam. An example was orders were taking longer to fulfill, and in some worst-case scenarios, customers and orders were mixed up. There was a severe inventory management issue that occurred in the holiday season in 2011.

When the company selected the ERP tool and consulting services, WWE was led to believe that the implementation was relatively easy. The truth could not be any different. In short, after working on the system for almost 18 months, the ERP launched on September 30, 2011. Immediately, there was a series of near-catastrophic errors – inventory issues, customer records mixed up, accounting errors, employee records went missing, etc. By late December 2011, in its busiest season, the CEO, Mr. Zhang, made the tough decision of reverting to the former set of systems (luckily the firm kept the new and old systems in parallel operations just in case of a disaster) and ended its first attempt at implementing the ERP solution. (Consequently, it sued the ERP firm and eventually settled out of court. The internal stories suggested that the settlement was hugely in favor of WWE.) Nonetheless, the company suffered immeasurably. Due to the delays, miscalculations, and customer service issues, the revenue for the 2011 holiday season was actually below 2010 revenue, and it was the first major setback for the firm. The stock dropped nearly 30% from its high in 2010, and the Board demanded immediately management's attention.

To be fair to the upper management team, they were already doing everything they knew at the time to stem the losses and resolve the problems. But the optics was bad even though the situation stabilized after terminating the ERP solution. Under Mr. Zhang's leadership, the senior management with the support of the Board agreed to create a new role in February 2012 – chief project officer (CPO) for the company at the enterprise level. The CPO's primary responsibility is the successful execution of strategic initiatives. WWE hired Korn Ferry and conducted a worldwide executive search. After 6 months of intense search and selecting from a pool of over 300 candidates and a shortlist of 15 highly qualified business executives, Mr. Zhang and the Board agreed to hire Ms. Adriana Holmes in August 2012.

Ms. Holmes has a long list of successes as a CEO of a smaller regional retail chain in Massachusetts. Through acquisition, she went to become a chief operating officer (COO) for a significantly larger retailer in the Northeast where she was responsible for a successful turnaround of the organization's operations and sales from being an industry laggard to a leader. In her spare time, she became heavily involved in a number of nonprofit organizations as a Board member, taught in some local schools, and became a champion of education and learning across the entire spectrum from K-12 to adult and lifelong learners.

What follows in the rest of this case study are the various projects and challenges that Ms. Holmes and her team had to confront.

D. People in the Case Study

There are four important people in the Global Information Store (GIS) project, and Table B.1 provides a brief description of these people.

Discussion Questions: If you are Mr. Zhang...

1. What is "total project success" for this project? Describe them in terms of "project" success and "project management" success.
2. Knowing what you know now, what would you have done differently in 2010 to better prepare the organization for the ERP implementation?
3. Is the concept "Duty of Care" applicable to this project? Why or why not?
4. What skills, competencies, behavior, and attitude should an ideal project manager have for this project?

Table B.1 Key People in the Case

Name	*Role*	*Description*
Darren (Wai Ming) Zhang	Chief executive officer (CEO)	Mr. Zhang is the founder and CEO of the firm. He is well respected and liked by all. As a consummate professional, he is courteous, respectful, and courageous. With an engineering background, he encourages experimentation, calculated risk taking, and systems thinking. Perhaps his biggest weaknesses are his stubbornness, occasional distrust of consultants and industry best practices, and impatience with overanalysis.
Adriana Holmes	Chief project officer	Ms. Holmes started with the firm in September 2009. She is a "self-made" executive, being the CEO of her firm, and became the chief operating officer of a much larger retailer (through acquisition). Even though she did not have a formal education or certification in project management, Ms. Holmes has built an incredible reputation as a great project leader. She prides herself on the ability to learn, and as such, she is heavily vested in the education industry.
Ben Alvarez	Head of PMO	Hired by Ms. Holmes in late 2009, Ben has more than 20 years of experience building and managing project management offices. He is certified in Project Management Professional, and he is planning to pursue additional certifications such as the Program Management Certification.
You	Project manager	In December 2013, Ms. Holmes asked you to lead the Global Information Store Phase 1 project.

Additional/Optional Questions of Considerations

5. Given the major failure of the ERP implementation, should the board ask you, the CEO, to resign? Why or why not?
6. Should you fire one or more senior executives for the project failure? Why or why not?

7. Would you have done the same – creating the role of CPO – or would you do something different? Please explain.
8. Does Ms. Adriana Holmes have the qualifications? What attributes would strengthen her candidacy?
9. Describe the most important, complex, or challenging projects that you have worked on? Use Template 1: Identifying and Analyzing Projects on page 263 in Appendix A.
10. Describe another project in which there is a clear violation of the "Duty of Care" guideline? Explain the breach of this duty.

II. State of Project Management

When Ms. Holmes arrived at WWE in September 2012, the timing was both fortunate and unfortunate. It was fortunate because the company is now reconstituting the ERP project, and she started at the very beginning of the planning. It was unfortunate because of the historical failure leaving no room for mistakes. Ms. Holmes, even before having a chance to bring in additional help, decided to conduct a high-level review of the state of project management in the company.

Through primarily by interviews with internal project leaders who worked on the failed ERP project as well as other project sponsors and managers, she learned the following:

- There was no standard methodology or approach for project management, from project selection, through project approval, through implementation, and closure
- There was little consistency of tools and processes. Of the three major projects that Ms. Holmes reviewed, only two tools were consistently applied: e-mail was the most popular communication and knowledge management tool, and nearly all project documents are literally dumped into multiple shared folders in SharePoint. Even the project status reporting had three different ways of gauging project performance.
- There were a few certified project managers in the organization, but their voice was largely muted. From speaking with many others on the project team, project management was largely viewed as bureaucratic (e.g., so many processes and firms and not sure of their value) and a lower priority than doing the actual work (e.g., the teams were so busy doing the work that there was little time for planning).

Discussion Questions: If you are Ms. Holmes…

1. Given what's stated here and in Section I, how well did the project follow the project management principles?
2. What do you believe are the five biggest challenges for this project, organized by knowledge domain?

3. Was the project too big and too long? Should the project manager consider adopting an adaptive methodology or utilizing a more traditional method? Support your reason.
4. How would you tailor the project management method? What would you emphasize as the most crucial aspects of the project?

Additional/Optional Questions of Considerations

5. As Ms. Holmes, what would you do in the next 100 days?
6. Would you stop the ERP project planning, accelerate the discussion, or maintain the current course? Why or why not?
7. After seeing all the challenges combined with an outsized expectation that you can turn the situation around, are you excited to tackle the challenges or are you contemplating resigning and leaving behind this big and messy situation?
8. It is clear that you will need help and that Mr. Zhang is expecting you to build a strong project management team, which skill sets do you believe are the most important? Most urgent?
9. For a complex and large project, there can be many additional knowledge domains. Based on your business experience, what other knowledge domain(s) may be crucial for the success of the project.
10. Describe whether you can apply the four guidelines of right sizing project management to this project? If yes, how.

III. Ideation Phase

During the chaos of December 2011, even though there was no real discussion of the board considering Mr. Zhang's resignation, Mr. Zhang nonetheless thought about resigning because of the ongoing fiasco and poor sales performance. Over the 2011 year-end holidays, Mr. Zhang reflected back on how he and his senior executive came to select the ERP.

With the rapid growth of WWE, the business was confronting significant business process and IT automation issues since 2007. A team of internal business and information technology experts was assembled to examine the situation and make recommendations in 2008. Some smaller ideas were approved and implemented, but those projects were largely limited in addressing the bigger challenges. Most of the ideas, however, were rejected as patchwork or unambitious. The company board eventually involved in the strategizing, and they wanted something bolder. At the same time, with each rejection, the team became both more frustrated and also more aggressive. By mid-2009 at the urging of Mr. Zhang and the board, the team came up with three competing projects for consideration, as shown in the table below:

#	Name	Description	Cost/Benefit	Duration
1	Enhancement	This was conceived as a super project with about ten specific projects within it. The goal was to improve the business processes and technology incrementally to keep up with the growth. This was essentially the continuation of what was started in 2008, and progress was good. But some of senior management and board wanted the organization to move faster, as the company's ability to innovate was hindered by process and technology.	$8.5 million annually of enhancements for the project. There was no net new benefit as the project is designed to keep the business operationally viable.	Foreseeable future
2	Enterprise resource planning (ERP) (modular)	The team considered implementing an ERP software, and they short-listed SAP or Oracle. The goal was to implement a limited number of major business processes per iteration, and each iteration is about a year. To implement all the known changes, it would take at least 5 years and possibly as long as 8 years. With every new process, the company can introduce innovations and changes with respect to that set of processes.	$15 million annually. The benefit is a significant improvement in capabilities, but modularly over the next 5–8 years.	5–8 years
3	ERP (accelerated)	Similar to #2, this proposal is to accelerate change. Instead of introducing new capabilities modularly, the first phase of the project will be to build most of the high-priority processes and systems and deliver them in 18 months or less. The subsequent phases will probably be more incremental improvements, but the details have not been discussed.	$45 million in 18 months for Phase 1. The benefit is accelerated benefit attainment.	1.5 years

Since the company was aggressively growing and had not experienced any major failures, the board and senior management made the bold choice of approving #3. But this was not without controversy. Two executives, the chief operating officer and chief information officer both recommended either #2 (COO) or #1 (CIO). But they were overruled by the chief executive officer, chief marketing officer, chief innovation officer, and eventually the Board. Two executives, head of human resources (HR) and head of business strategy, abstained. Now in hindsight, Mr. Zhang thought whether there should be more evaluation of the other proposals and discussion to better understand the other executive's concerns. To Mr. Zhang, there were two major puzzles after approving #3.

1. The business case was clearly wrong – underestimating cost and ease of implementation. At the same time, the benefit of better reporting and analytics was overstated. The COO raised this concern at the time, but since there was no clear counter solution, it was rejected.
2. Everything took longer to accomplish. For example, overstating benefits underestimated the implementation schedule. For example, the business requirements were allocated 3 months to complete. But by month 6, the team was still struggling with getting executive buy-ins. The cost outpaced the plan by nearly 50% by the fifth month. CIO and head of HR raised the concern of the implementation schedule, but it was rejected by the general enthusiasm of Mr. Zhang and the Board.

Discussion Questions: If you are Mr. Zhang…

1. Were these projects (#1–#3) aligned with the company strategy? Why or why not? Support your answer with evidence from the case written to date?
2. Should the organization's project management capability in 2010 be a factor in business case? Why or why not?
3. How about culture? How does culture impact the implementation of the chosen project?
4. What additional factors (which may not be listed above) should the executives and board consider before making a decision?

Additional/Optional Questions of Considerations

5. Why was the business case so wrong? Based on your working experience, what were some potential reasons for the poor business case?
6. Should the company select either #1 or #2? In hindsight, was selecting #3 the best option? Why or why not?
7. Could a tool like analytical hierarchy processing help with project selection?
8. If you were on the team making a set of recommendations, what would be your recommendations to rescue the project? Describe up to three recommendations.

9. If the organization had a project management office at the time of Ideation of this project, what may be some ways in which the PMO could improve the likelihood of success?
10. If you were Mr. Zhang and had the opportunity to restart the project, how would you have done?

IV. Initiation Phase

In addition to evaluating the events during the Ideation phase, Mr. Zhang also recalled the early days of the project. After the executives approved the project in early 2010, Mr. Zhang appointed a senior project manager to lead the project. But in the rush to start the project and believing the key project parameters were clear, Mr. Zhang and others pushed the project team to make tangible progress quickly and skip some key project management deliverables including the project charter.

Discussion Questions: If you are the project manager assigned to the project…

1. Given the size of this project as the executives approved option 3 and knowing what you know now, what would be the key Initiation activities that you would have undertaken?
2. If the executives asked you to skip some Initiation activities such as creating the project charter, what would you have done to resist the request?
3. What were the challenges of getting started on an ERP project of this kind? Refer to Wikipedia's description of "enterprise resource planning" to become familiar with this concept.[1]
4. Did the challenge of "fuzzy front end" impact this project? If yes, explain the type and degree of impact. If no, explain why no.

Additional/Optional Questions of Considerations

5. Identify one key question to address for each of the knowledge domains.
6. Rank-order the questions, from the most impactful or urgent to the least.
7. What are the key project and organizational factors that contributed to the project challenges? Identify two of each:
   a. Project factor
   b. Organization factor
8. On the environmental factors, identify categories 4–6 factors that contributed to the complexity of the project.

[1] Enterprise Resource Planning, (n.d.) in *Wikipedia*. Retrieved from https://en.wikipedia.org/wiki/Enterprise_resource_planning on November 20, 2019.

9. Review the list of ten good practices for getting started in Table 4.2: Good Practices for Getting Started on page 63, discuss three factors that you believe to be the most relevant.
10. Critically review the seven steps to managing the fuzzy front end described in Table 4.3: Good Practices to Manage Fuzzy Front End on page 64. Are these seven steps sufficient and appropriate? If not, please suggest additional steps.

V. Preparation Phase

Continuing from the earlier case, assume that Mr. Zhang and the senior executives accepted your recommendations to develop the project charter and other project initiation deliverables from Section IV. Even with a considerable project management effort during the Initiation Phase, the problem of getting started or the fuzzy front end was still considerable. Thankfully because of the planning and the executive support, there were no major obstacles as your team exited the Initiation Phase.

After recovering from the small battle scars from the Initiation Phase and energized by the support and moment of the relative success of completing the Initiation Phase, you as the project manager understand that the Preparation Phase will be critical for project success.

Discussion Questions: If you are the project manager assigned to the project…

1. Based on your understanding of the project as described from Section I to current, select three knowledge domains that you believe are the most important to focus early in the Preparation Phase. Describe why you believe these knowledge domains are the most important.
2. Examine each of the three domains carefully and determine the following:
   a. Key questions to address for each knowledge domain
   b. Risks associated with the selected domain that may negatively or positive impact the project
   c. Domain-specific project challenges that may have already impacted the project

Additional/Optional Questions of Considerations

3. Consider the other knowledge domains that are relevant and applicable to this project. Ask the same questions as in 2a, 2b, and 2c.
4. Examine the complexity of the project from the three perspectives of human behavior, system behavior, and ambiguity. Develop examples of each.
5. Develop recommendations on how to address these complex factors.

VI. Implementation Phase

As Mr. Zhang reflected on the past experiences, even though he readily agreed that there was insufficient planning and organizing at the Ideation, Initiation, and Preparation Phases from March through June 2010, the problem really started in the Implementation Phase. Perhaps, it was a lack of adequate project management preparation that led to all the problems, but he still shuddered remembering the many missed deadlines and complications.

For example, he vividly recalled a management meeting with his senior staff when the initial testing results failed to meet expectation. The Quality Assurance team reported a passing rate of 90% across the 2,000 test scripts, most of which was automated. What was detrimental was that of the 200 failed scripts, nearly 25% of them were in high severity categories – meaning that that the system would stop functioning and manual workaround must be found. The team recommended an additional round of system testing to improve the testing before moving into the next phase of testing with users. But it would also slip the schedule by nearly a month. Mr. Zhang, eager to push ahead, urged the team to reject the extra testing and head into the next phase of testing. Mr. Zhang thought that any additional errors could be found and addressed there. With great hesitancy, the other executives followed his lead. To his great relief, the next phase of testing with the users proceeded better than the expected. But shortly after the launch, the problems discovered earlier came back to haunt the new ERP system. In hindsight, the user testing was probably inadequate and failed to capture some key problems. At the worst moment, which occurred in week 4 after launch in October 2011, the incident management system was registered over 100 incidents per hour, completely overwhelming the team's ability to even review them. After 3 months of constant fires, resulting in some major errors on customer orders and departure of several high-performing employees, Mr. Zhang eventually ordered to restart the old processes and systems.

Here are some additional articles of consideration as you complete the discussion questions:

1. Fruhlinger, J. & Wailgum, T., 2019. "15 famous ERP disasters, dustups and disappointments". *CIO*. Retrieved from www.cio.com/article/2429865/enterprise-resource-planning-10-famous-erp-disasters-dustups-and-disappointments.html on October 29, 2019.
2. Schiff, J.L., 2017. "11 common ERP mistakes and how to avoid them". *CIO*. Retrieved from www.cio.com/article/2397802/article.html on October 29, 2019.

Discussion Questions: If you are the project manager assigned to the project…

1. Based on your understanding, what are the most important roles and responsibilities of the project manager on this project during the Implementation Phase?

2. Clearly the project was not proceeding well. What approach would you have taken to address the fundamental problems with the project?
3. In the meeting in which Mr. Zhang pushed ahead and rejected your recommendation to take time for another round of testing, you were clearly dismayed. What would you have done after the meeting?
4. In hindsight, what would you have done before this crucial meeting to improve the chance that your recommendation for another round of system testing would be accepted? (Hint: Review Section 6.7 on page 88.)

Additional/Optional Questions of Considerations

5. What are the biggest challenges during the Implementation Phase?
6. On ERP projects, what do you believe to be the biggest mistake that project managers make during the Implementation Phase?
7. As the project manager, you know the project is not going well. But the project sponsor and executives are not listening. What are some of the tactics you can use to bring greater credibility to your warnings?
8. Knowing what you know now, what would you have done differently if you had the opportunity to restart the project?

VII. Closure Phase

According to the original project, after the launch of the project on September 30, 2011, the project team would work closely with the newly formed Support Team for a period of 3 months of intense “hypercare”. This would ensure that incidents could be addressed quickly and also provide the Support Team with a period of learning of the new system. The project team could then formally close the project by year end and move on to the next project.

Sadly, the project did not proceed as planned. By February 2012, even the Board was involved as the situation deteriorated. This eventually led to the creation of a new position, the CPO, with the responsibility of fixing the situation.

The time was February 2011, 5 months after go live. The situation was getting worse in all areas:

- More customers’ errors, which is now happening regularly. An estimate shows about 2% of the errors contained mistakes, mainly in the invoices (price and quantity), shipping address (which led to a number of high-profile mistakes), and wrong customers. Worse, there was no clear pattern of errors; they seemed to come from everywhere.
- New system problems were still being found, even though the back log of defects was trending lower. Analysis shows that some errors were fixed, then broken by subsequent repairs, and so on. In a few cases, the cycle of repetitive breaking and fixing became a routine.

- Nearly everyone involved lost confidence in the new system. For high-profile customers, customer representatives were double and triple invoices and records manually, causing a significant jump in overtime pay, especially for January and February, which traditionally are the low seasons.

Discussion Questions: If you are the project manager assigned to the project…

1. Mr. Zhang asked for your thoughts on the potential of closing the project versus keep going to fix the system. What are your recommendations and the pros and cons of each?
2. If your recommendation was to close the project prematurely, what would you do to ensure a methodical shutdown and preserve as much of the valuable assets as possible?
3. If your recommendation was to plow ahead and fix the system, what would you have done to complete the system and rebuild the confidence that the system can be fixed?
4. Assuming you are a competent project manager and made good suggestions along the journey, but many of the good suggestions were rejected by Mr. Zhang and others, how would you feel? Would you quit? (Please note that Mr. Zhang and the other executives recognized their mistakes in key areas, so they are not assigning blame to you… at least not that you know of.)

Additional/Optional Questions of Considerations

5. If your recommendation is to close the project prematurely, should the team perform a post project evaluation? Why or why not?
6. If your recommendation is to close the project prematurely, what other closure activities should you consider?
7. In hindsight, it was clear that the project should never be launched in the first place. Based on your knowledge of the case, the book, and the suggested readings, how can you convince the executives to delay the launch?
8. In the final analysis of the project, what are the top factors that contributed to the failure of the projects?

VIII. Integration Management

About Global Information Store

Introduction

In the relaunch of the ERP system, now with Ms. Adriana Holmes as the new CPO, the project team reexamined and refreshed the business case. After a detailed analysis, the team agreed to apply a program management approach to manage the overall

ERP implementation. Within the ERP program, however, there were a number of important specific projects to consider including the sequencing of them.

After much deliberation, it was recognized that one of the previous biggest challenges was the inconsistency in the data design from the earlier attempt that led to so much confusion. Even though the actual ERP application is still under investigation, everyone eventually agreed that the company needed to launch a project to build a master data management system (MDM) that can establish and enforce clarity of important data elements across its customers, products, and people. The GIS project was launched in December 2012 to build the MDM system.

Note to reader: To learn more about MDM, refer to a list of credible resources in Appendix D of the case study.

1. Wikipedia is a good source to start the understanding of MDM: https://en.wikipedia.org/wiki/Master_data_management
2. This Wall Street Journal article, sponsored by Deloitte, provides a summary business case for MDM. http://deloitte.wsj.com/cio/2013/07/24/the-business-case-for-master-data-management/
3. Even though this article positions the Microsoft solution, it provides valuable information on the "what, why, and how of MDM". https://msdn.microsoft.com/en-us/library/bb190163.aspx

The first phase of the GIS system includes the three most important types of data: People, Products, and Customers. The GIS People brings clarity to how the organization and its systems view and manage its employees and contractors. This role-based system will eventually be used by other systems, including the ERP, for authentication and access control. Plus, in a future phase of the ERP, this GIS People will be an invaluable input to the HR system. The GIS Products formalizes the product and services categories of the 20 million (and growing) SKUs throughout the organization. By standardizing the product and service categories, the organization can assign proper product management processes including service delivery. GIS Customers is required for the ERP system to better manage the supply chain. It also has a strategic importance for another project under consideration – customer relationship management (CRM) in which the GIS Customers will be an important foundational component. There are additional GIS components being planned for future phases. This includes WWE retail store, distribution center, and vendors and suppliers.

After much deliberation, the project team agreed with these strategic business drivers for the GIS system:

- **Single source of truth.** GIS system will be the system of record for the information in its system.
- **Reduce redundant data.** GIS system will be the only system for these records, to avoid confusion.

- **Single point of accountability.** Once done, the GIS system will have one global team to manage the system and govern the processes for data identification, categorization, cleansing, use, and termination.
- **Continuous improvement.** This team will be responsible for the ongoing management of this application and its data as a strategic asset of the organization.

Phase 1 of the GIS System is estimated to be $10 million and requiring about 1.5 years to complete, after the sign-off of the project charter. Given the importance of the project, Ms. Holmes will be the project sponsor, and she will also be forming a Governance Committee in which Mr. Zhang will also be participating. The ongoing operating budget for maintaining and improving the system is estimated at $2 million per year. Both the budget and schedule are approximations, and refinements are expected. However, given the history, the need to produce a quality system is the highest priority. Mr. Zhang has also suggested that WWE should use this project as a way to develop its employees.

One of the noticeable failures of the previous attempt with the ERP implementation was treating the project as an information technology project. As such, the previous project team, including Mr. Zhang, was viewing the project through the lens of technology challenges – how to make the systems talk to each other. But ultimately, the failure was not so much that the systems did not "talk" with each other, it was that the systems were "talking over" each other. Furthermore, there were significant employee issues – not knowing why the system behaves the way it did or why some tasks that were very simple in the legacy system now take three times longer to perform. Apparently, there was also poor alignment between the system and the business processes, and in the second week after the launch, there were 2,500 incident tickets generated. The project and the operational teams were so overwhelmed, as they were designed to handle just 100 tickets per day. Thus, many incidents "fell through the cracks" and the resulting problems impacted customer service and delivery.

Learning from past lessons, Ms. Holmes has insisted on better integration management. Working with the project manager for the GIS System, the team agreed to implement the following processes:

1. **Integrated change management.** All changes will be deliberately managed after their approval.
2. **Dependency management.** Project manager is responsible for overseeing the dependencies across multiple teams within the project.
3. **Reports.** Detailed project performance reports will be shared across the entire project team. All team leads and above are required to read them.
4. **Meetings and communication.** Project meetings and broader communication are planned in advance with mandatory attendance of every manager and above, of impacted departments.

Ms. Holmes and the project manager believe that through these activities conducted regularly throughout the project, the GIS project will be able to achieve a strong degree of integration across the various project teams within the overall project.

Timeframe for the enterprise resource planning and Global Information Store Projects

Table B.2 shows the major timeframe for the ERP and GIS projects.

Discussion Questions

1. Why is project integration important?
2. Do you agree with the processes described in Chapter 8 to achieve sufficient integration? Why or why not?
3. What are some other activities you would recommend to ensure better integration?
4. Based on your understanding of the project, what are some areas that are more susceptible to change?

Additional/Optional Questions of Considerations

Congratulations. Right before the official launch, Ms. Holmes has assigned you to be the project manager leading the GIS System Phase 1 project.

5. Will you implement a less or a more rigorous change management process? What are the pros and cons of each? (Hint: You can also suggest a tiered process in which the rigor depends on the type of change)
6. How do you manage the project knowledge that would be created on this project?
7. As this project is a part of the ERP program, how would you coordinate the GIS project activities with the overall ERP program and the other project managers within the program?
8. Since Adriana Holmes is new to the organization and you have never worked with her, what would you do to build a positive and constructive relationship with your project sponsor?

Exercise

1. Create a project vision for this project. Use Template 2: Project Charter on page 265 in Appendix A.
2. Your first job is to create a project charter and obtain approval from the executives. Based on the information given and using the project template, create a project charter for this project. Consider using Template 2: Project Charter on page 265 in Appendix A.

Table B.2 Timeframe for the Case

Date	*Event*	*Description*
March 2010	Relaunched enterprise resource planning (ERP) implementation	To manage growth and to integrate the company into one system, the company launched a major ERP initiative in March 2006.
June 2010	Mr. Zhang's preparation for the town hall	Up to this point, the project performance reports indicated that the project was healthy. But as Mr. Zhang was preparing a town hall meeting with the employees to prepare them for the ERP systems launch, he was shocked at the poor responses to his questions. This is the first inkling that something major is wrong. He immediately allocated $500k extra per month to keep the legacy systems in operation for as long as needed for the new ERP system to stabilize.
September 30, 2011–late December 2011	Launch of the new system	The system launched on time, but it immediately faced unexpected challenges and errors. Thankfully, the legacy system remained in operation, and every night, the analysts reconciled the various data points between the legacy and the new ERP system. The expectation was that the number of errors would rapidly decline as errors are identified and fixed. Unfortunately, the error rate did not decline, and it affected the overall operations. Even though the legacy system was in operations, visible mistakes still occurred because everyone was so busy. With no end in sight, Mr. Zhang terminated the new ERP system in early 2012.

(Continued)

Table B.2 (*Continued*) Timeframe for the Case

Date	*Event*	*Description*
February 2011	CEO and Board agreed to create CPO	Given the catastrophic failures, WWE executives and the board agreed to create a new executive officer – the chief project officer (CPO) to oversee all enterprise projects.
September 2011	Ms. Adriana Holmes started	Ms. Holmes started as the CPO.
October–December 2012	Global Information Store (GIS) project ideation	The main goal of this project phase is to reapprove the implementation of the ERP project, in particular the GIS project.
December 2012	You are assigned as the project manager	Your first job is to create a project charter, which was completed in February 2013.
December 2012	Start GIS system project	Shortly after Ms. Holmes started at WWE, the executives agreed to restart the ERP systems planning. From the lessons learned, everyone also recognized that for the ERP to work, a master data management application is going to play a crucial role – the GIS.
December 2012–March 2013	Project Initiation	The main goal of this project phase is to work on the project charter collaboratively with the key stakeholders and also to obtain approval of the charter.

(*Continued*)

Table B.2 (*Continued*) Timeframe for the Case

Date	*Event*	*Description*
March–May 2013	Project Preparation	In this phase of the project, the focus turns to the detailed planning of the project and prepares it for execution. Key deliverables include the project management plan and other detailed project planning artifacts.
May–December 2013	Project Implementation	This phase is potentially longer than all the other phases combined, and the focus is to build a working GIS.
December 2013–March 2014	Project pilot and training	This is arguably a subphase within the Project Implementation, and the purpose is to pilot the system in certain countries before it is launched into general production. Also training for the new system is planned for this period.
April 2014	Launch and Transition	This is the final phase of the project in which the primary project deliverables are rolled out to the entire organization. There is a planned buffer of up to 2 months in which the project team will provide support before transitioning to work on the next phase of this project.

3. As the project manager for the GIS System, create an integrated change control plan for the project.
4. What does project success look like to you? Create a list of attributes, preferably with specific measures.

IX. Stakeholder Management

With the project charter approved by February 2013, the GIS System is moving into the Preparation Phase. You as the project manager just had a lengthy meeting with Ms. Holmes and Ben Alvarez, the newly hired head of the project management office (PMO) and discussed some of the biggest failures from the earlier attempt on the ERP system. These include the following:

- WWE executives failed to understand the difficulty of ERP implementation. Starting with Mr. Zhang, the entire management team failed to understand the complexities and challenges of the ERP implementation. This was partly because there was no prior experience of ERP implementation and partly because the ERP vendor painted a rosy picture. Nonetheless, the management team criticized itself heavily for the lack of executive attention.
- ERP vendor was poorly managed. The internal project manager accepted the performance reports without much questioning. In preparation for the lawsuit, it was discovered that much of the performance reporting was based on overly optimistic interpretation of progress, with borderline deception.
- Project stakeholders were not engaged. In a hurry to launch the project and believing that the business case was obvious, the project team started doing work without understanding the impact to all parties – internal employees; suppliers and vendors; customers both retail and institutional; and various shareholders. This led to poor expectation management, and once problems started, there were significant numbers of conflicts.
- There were many other challenges, but the above three were the most relevant to your next assignment.

In hindsight, what saved the company from even a bigger fiasco was fortuitous. In preparation for a town hall meeting in May 2010 (about 2 months after the launch), Mr. Zhang pushed the project manager and the ERP vendor for specific capabilities and readiness of their systems, and the answers received were ambiguous. They promised better order management, for example, but were not able to explain what "better" meant. On training for the new system, the ERP vendor presented a credible plan but also missed some important business process considerations, which meant that the operator would know how to use the system but not understand "why". In short, Mr. Zhang felt so uncomfortable with the meeting that he went back to his management team and raised serious alarms. By then, it was too late to fix much of the concerns, and the management team agreed to invest $500k per month extra to keep the legacy system in operation for as long as needed until the new ERP system proved its working capabilities. Mr. Zhang further insisted that the legacy and the new system reconcile its differences every night until the differences become negligible. (Sadly, the differences were almost random, leading Mr. Zhang to shut down the new system in March 2011.)

As one of the first steps in project planning, Ms. Holmes, Mr. Alvarez, and you agree that conducting a stakeholder analysis is incredibly important.

Discussion Questions

1. Why is stakeholder analysis important (or not important)?
2. In your view, who are the top three stakeholder groups?
3. Using the Power/Interest Matrix, how would you categorize these stakeholder groups?
4. How to monitor the effectiveness of stakeholder engagement?

Additional/Optional Questions of Considerations

5. This project has many stakeholders, likely over 1,000 throughout the company. What is your plan to manage them?
6. Given the lessons learned, what are the top three things that you as the project manager would do differently on this project (aside from the stakeholder analysis)? Please explain.
7. What are some of the biggest challenges with stakeholder engagement?
8. During the project implementation, suddenly you recognize a new stakeholder group gained prominence. They were previously evaluated with a low score, but the importance gained significantly with a routine reorganization. How do you manage this change?

Exercise

1. Stakeholder evaluation matrix. Using the stakeholder evaluation matrix in Figure 1.1: Project Dimensions Beyond Scope, Cost and Time on page 8, perform a stakeholder evaluation for this project.
2. Stakeholder engagement plan. Based on the tool shown in Figure 9.3: Stakeholder Engagement Plan on page 120 and the output of the stakeholder evaluation matrix created above, develop a stakeholder engagement plan for this project.

X. Scope Management

As described in Section VIII of the case study, this current project, GIS Project Phase 1, focuses on creating an application and the associated processes to manage three sets of master data:

- GIS People brings clarity to how the organization and its systems view and manage its employees and contractors. This role-based system will eventually be used by other systems, including the ERP, for authentication and access control. Plus, in a future phase of the ERP, the GIS People will be an invaluable input to the HR system.
- GIS Products formalizes the product and services categories of the 20 million (and growing) SKUs throughout the organization. By standardizing the product and service categories, the organization can assign proper product management processes including service delivery.
- GIS Customers is required for the ERP system to manage the supply chain better. It also has a strategic importance for another project under consideration – CRM in which the GIS Customers will be an important foundational component.

For each of these set of master data, the project team plans to create the following project artifacts:

1. Vision for this set of master data (e.g., customer, product, people)
2. Business requirement
3. Technical specifications and engineering diagrams
4. Communication materials
5. Training content
6. Operational guide including detailed process description
7. Enhancement backlog including tracking of issues and changes

There are additional GIS components being planned for future phases. The important functions of the application and business processes are as follows:

1. Clear processes of identifying, defining, evaluating, approving, and terminating any new data. All data also must have clear ownership, preferably a single point of ownership.
2. The application must be able to create an auditable trail of all changes including the following:
   a. Importing or adding new data into the application
   b. Modifying and deleting existing data
   c. Cleaning data, especially updating with updated information

   d. Merging data, especially when the same entity has multiple entries
   e. Tracking data usage by other systems and processes
   f. Maintaining a "hot" backup in the case of a system disaster
3. In addition to the processes and applications, it is incumbent upon this project to create an organization adoption project to educate all managers and above, employees of WWE, and training programs for all administrators and superusers.

Discussion Questions

1. Based on the above, write a high-level scope statement.
2. Discuss the top five to ten major project activities required to complete this project.
3. What are the major challenges that you can envision when it comes to achieving agreements across those three categories of data? Are there conflicts or politics?
4. If you are the project manager, would you want to limit the initial project scope to just one master data type (e.g., people, product, or customer) or do you believe the project can handle more than three? Why or why not?

Additional/Optional Questions of Considerations

5. What approach would you take to define the master data? If the approach differs across the three types, then explain.
6. Across the entirety of the WWE, there are multiple information technology teams that must agree to the adoption and utilization of GIS. For many IT teams managing various applications and systems, GIS will be more than a process change. They would need to change the business processes for managing people, customer, and products too. What are some organizational issues that you are likely to counter?
7. Should the scope of work required to manage the organization change to achieve good adoption be a part of this project? Why or why not?
8. In the GIS application roadmap, the next phase of inclusion would be creating the master data for the supply chain management (SCM). Would the knowledge of this next phase impact the current scope of this project? Why or why not?

Exercise

Based on the description in Section 10.3 Planning and Defining Scope on page 128, conduct the activity planning exercise and create a high-level work breakdown structure for this project.

XI. Schedule Management

The GIS Project Phase 1 project charter was approved in February 2013. It was estimated that the detailed planning, execution, and closure will take an additional year to complete. Working closely with Ms. Holmes and Mr. Alvarez and also with the project team leads, the project manager believes that 1 year is feasible. The major phases of the project with the start and completion dates are as follows:

1. Project Ideation (main deliverable: re-approval of ERP and GIS)" Oct 2012–Dec 2012
2. Project Initiation (main deliverable: project charter): December 2012–March 2013
3. Project Preparation (main deliverables: project management plan, detailed project planning artifacts): March 2013–May 2013
4. Project Implementation (main deliverables: application, business processes): May 2013–December 2013
5. Project Pilot and Training (main deliverables: working system that can be piloted, training for operations team): December 2013–March 2014
6. Launch (main deliverables: successful launch and transition to operations team) – April 2014

Given the importance of the project, Ms. Holmes and the Governance Team also agreed to two additional months as a buffer. They envision, by March 2014, the new ERP project will be well underway, but the integration with the new ERP project is planned for May 2014.

Discussion Questions

1. The project schedule above is organized by life cycle. Do you agree with the six phases listed above? Should there be more or fewer phases? If you have free rein as the project manager to create the phases, what would you choose? Please explain.
2. Identify three project scheduling challenges and explain how you would address these challenges.
3. From a project sequencing perspective, which activities should be identified first: business processes for managing the data or the business and functional requirements of the application? Why?
4. Do you believe a phase for piloting is important for a project like this? Why or why not?

Additional/Optional Questions of Considerations

5. Ms. Holmes and the Governance Team agreed to two additional months of project schedule contingency. But would it be better to include the contingency at the task level instead of adding the additional time at the end of the planned schedule? What are the pros and cons of each?
6. Under what situations would make this buffer of 2 months insufficient or at least uncomfortable?
7. Of the six phases of the project, which phase is likely to be the most resource intensive? Why?
8. As the project manager, you expressed concern that the timeframe is very tight. Discuss two ways of managing the situation to reduce your ease?

Exercise

Based on the project activities identified in Section VIII and combined with more information above and your analysis, create an integrated project schedule of about 80 tasks for this project. Remember to select any of the project activity and drill down to a detailed level.

XII. Resource Management

As the project manager plans and prepares the project activities, the project is clearly more complex than originally planned. After discussing the project with the core team leads, it was generally agreed that the project team considers the following types of resources:

- Project manager
- Business analyst
- Technical architect
- Technical engineer
- Developer
- Testers
- Trainers
- Deployment specialist
- Communication specialist

As project stakeholders of this project span across the globe, the project manager is aware of the high need for team players with strong interpersonal skills. In some situations, having the right personality may be more important than having the technical skills.

Discussion

1. Given what you have learned of WWE and this project, identify five soft or technical skills that are important for the project manager.
2. Is it more important to find someone who has managed ERP and/or master data management projects versus having the soft skills to tackle the organization challenges?
3. If you are the project manager, what do you look for in a sponsor? Explain.
4. Should the project manager strive to hire people who can work from Saratoga Springs where most of the project team reside, or can team members reside anywhere in the world? Please explain.

Additional/Optional Questions of Considerations

5. How can the project manager best motivate the core team members?
6. What are the important factors of consideration in the estimation of the quantity of project resources required?
7. In addition to HR, what other resources are required for the successful delivery of this project?
8. Based on your understanding, discuss what you believe to be the most difficult resource (human or nonhuman) to secure for this project.

Exercise

1. Enhance the integrated project schedule created in Section IX of the case study by adding the role names.
2. Create an RACI matrix for the key artifacts discussed in Section X of the case study and the roles listed in this section of the case study.

XIII. Cost Management

When the GIS Project was first conceived, the project team at the time came up with the US $10 million budget and 15 months for implementation using a combination of analogous estimation techniques and expert opinion. But now with the project in full swing and the project preparation underway, the executives would like to have a more accurate project budget.

Discussion Questions: As the Project Manager…

1. What are some techniques that can provide a more accurate project budget? Discuss the pros and cons of each.
2. Which technique would you choose to develop a more accurate budget, assuming you are in Project Preparation Phase? Explain why you select this technique.

3. Identify three challenges that prevent you from developing a more accurate budget?
4. Which types of project resources are likely to be the costliest for the project? Why?

Additional/Optional Questions of Considerations

5. If the executives ask you to cut the project budget by 10%, how would you plan for the cut?
6. If the project executives provide you with 10% more budget than requested (given the importance of the project), how would you spend the extra 10%?
7. Is earned value management an appropriate method of managing project performance, in particular cost on this project? Why or why not?
8. Which of the budget scenario would you rather confront? Explain.
   a. The estimation of the various tasks and activities is inaccurate (e.g., off by as much as 30%), but it is relatively precise.
   b. Comparatively, the estimation of the various tasks and activities is more accurate than above (e.g., off only by 15%–20%) but much less precise.

Exercise

1. Create a project budget based on your understanding and clearly document the assumptions.
2. For question #8 above, graphically illustrate the two scenarios.

XIV. Project Communication Management

From a communication perspective, the postproject evaluation of the original ERP project also showed major concerns. Clearly, the project communication was ineffective. There were a lot of issues raised throughout the project, and groups were essentially talking "over" each other and not "with" each other. Furthermore, there were concerns with the timeliness and accuracy of the information. Even though the previous project manager raised the awareness of communication importance, the project executives thought little of the concerns and they were rarely followed through with credible actions that remedied the situations. (For these reasons, the management team blamed itself and did not blame the previous project manager.)

To summarize, three major groups of communication issues were raised:

1. **Timeliness.** Even though there should have been weekly performance reports, the information in those reports was infrequently updated. For example, the budget almost always lagged the current information by about a month.

The progress on activity completion was lagging the actual. Some concerns were raised, but band aids were applied, and there was no systemic change.

2. **Accuracy.** The ERP vendor and project manager painted a far more optimistic picture of conformance to the standard project performance metrics than was happening in reality. The previous internal project manager raised a series of concerns when there were disconnects between the reports and the actual deliverables. But the Management Team, including Mr. Zhang, was too busy dealing with other company activities to push the ERP vendor. Plus, the ERP vendor had an effective engagement manager that at least appeared to address any escalated issues. The previous project manager was so frustrated and felt disempowered that he left the company shortly after the launch (which created more internal turmoil).
3. **Chaotic management of information.** Since the ERP was the largest project of its kind, WWE did not have the tools and processes to manage the information. There was no single project management information system (PMIS), and there was general confusion, even among the project team members, on the location of the latest information. This problem is compounded by having about 1,000 stakeholders on this project.

There were other considerable issues too. Most of the project team worked from the Saratoga Springs office, even though there were offices and distribution centers around the world. Project meetings were poorly attended. Some remote project team members felt disconnected with the project.

Discussion Questions

1. Why is project communication important on this project?
2. Given the situation above, what would you, as the project manager, have done to improve project communication?
3. Should the company implement a PMIS for this project? Why or why not?
4. What are your recommendations to improve communication management with remote teams? What would you do to engage the remote team members more?

Additional/Optional Questions of Considerations

5. With about 1,000 stakeholders, how many potential communications are there?
6. How do you manage communication effectiveness with so many stakeholders and channels? Recommend three ways to manage communication on this project?
7. What metrics would you use to measure communication effectiveness? Develop two metrics and describe them.

8. WWE is a global organization with a diversity of people. Identify three communication challenges related to this diversity of people on this project?

Exercise

1. Leveraging the stakeholder analysis from Section IX of the case study and Figure 14.3: Example Communications Management Plan on page 181, create an effective communication plan.

XV. Project Risk Management

Given the failure of the previous attempt to implement the ERP system at WWE, the project experienced a high number of risks. Many of these risks happened due to significant issues for the project that eventually doomed the first attempt. But much of the same risk remains with this latest attempt to implement the system.

Now as the project manager, the good news is that you have ***access*** to the risk registry of the previous attempt, many of the same people who worked on the project before, the newly hired CPO, and the attention of the other C-level executives. How do you plan to leverage this wealth of information and support to mitigate threats and maximize opportunities for the current project?

Discussion Question: As the Project Manager…

1. What approach would you take to risk management?
2. How do you plan to identify the risks?
3. How would you evaluate the risk priorities? What are the key factors of consideration?
4. How do you plan to organize the risks? What are the likely risk categories to consider?

Additional/Optional Questions of Considerations

5. Are there positive risks or opportunities on this project? Please identify three opportunities and describe.
6. Are there negative risks or threats on this project? Please identify three threats and describe.
7. Since the GIS is a subproject within the overall ERP project, propose a process to manage risks between these projects?
8. Since the company has learned much from the previous experience, is risk management still important to this project? Why or why not?

Exercise

Using the Template 6: Sample Project Performance Dashboard on page 273 in Appendix A, identify up to eight risks, including both positive and negative risks, and analyze them using the template.

XVI. Quality Management

As the previous attempt shown, the quality of this system is extremely important to this project. Project quality does not occur by accident. Project managers work hard from the project initiation to make sure quality is planned into the very fabric of implementation. As described in the textbook, there are two important considerations: fit for purpose (utility) and fit for use (warranty). For the GIS Project, they are as follows:

I. **Fit for purpose (utility).** Refers to the ability of products/services to meet the intended needs. For GIS Project, each of the master data types serves as the definitive categorization of the data. The application is the repository for the specific data, which is used by all other information systems requiring People, Products and Services, and Customer data.
II. **Fit for use (warranty).** Refers to the performance and accessibility of these three types of data. It can also refer to the security, maintainability, scalability, and availability of the data. As this is a master data servicing all systems eventually, a particular concern is service disruption, for example. To address disaster recovery, there is a plan for active mirrored sites throughout the world in which the availability is essentially 100%. It would take a global catastrophe to shut down all backup sites around the world simultaneously.

As mentioned in Section III, WWE experienced about 17.5 orders per second or 1.5 million transactions per day. Each of these transactions has the potential to touch one or more of these data types.

Discussion Questions

1. Why is quality management important for this project?
2. Discuss a likely case in which a quality issue can lead to a customer problem.
3. Which is a more important aspect of quality: fit for purpose or fit for use?
4. If you are the project manager, what would you do to achieve the desired project quality?

Additional/Optional Questions of Considerations

5. What does quality mean for this project? How does the project manager determine quality for this project? Suggest an approach to determine quality and propose five quality metrics.
6. During the middle of the project's Implementation Phase, you found that while the project team was following the quality process to the *letter*, they were not embracing the importance of quality in *spirit*. This was especially evident in the test planning. Emphasis was placed on the common use cases, but every few people were thinking outside the box and come up with the more exceptional uses and cases. Highlight two good practices that you plan to implement or reimplement to emphasize the importance of quality.
7. During the project, your project sponsor asked you to consider an audit before the project is largely completed. You considered various options, formal versus informal or internal versus external. What do you recommend? Why?
8. On this project, who is ultimately accountable for project quality?

Exercise

Create a quality management planning document based on the following:

1. **Quality management process.** Describe how the project will achieve the defined quality.
   a. **Definition of quality.** Define what is quality for this project.
   b. **Quality metrics.** Highlight the five most important quality metrics (in your opinion).
   c. **Roles and responsibilities.** Highlight who (person or role) will be responsible and accountable.
   d. **Tools.** Implement quality management tools, in support of quality metrics.
2. **Quality assurance.** Ensure proactively quality is being achieved.
   a. **Assessment, reviews, and audits.** Identify project reviews, process audits, and other assessment that should be performed during the project life cycle.
   b. **Quality control.** Identify quality monitoring and intervention activities.
   c. **Corrective actions.** Describe the process of tracking, resolving, and reporting defects, anomalies, and other problems.

XVII. Supply Chain Management

One of the early planning activities was to determine whether WWE should build the product or purchase an existing product. After a thorough analysis of the available systems on the market, including master data systems that are a part of SAP,

Oracle, Microsoft, and other ERP solutions, the project team determined that it makes more business sense to build the system, instead of spending heavily to customize an existing product, pay the initial licensing fee and ongoing maintenance fee. WWE would still be responsible for compatibility issues with future upgrades. Strategically, utilizing an existing product also means too many compromises on ease of upgrade and performance.

Building the system possesses its own risks, starting with the software development capabilities. Luckily, on an unrelated project, WWE has developed a strong partnership with Centaura. After a cursory process of vetting competing vendors, Centaura was selected to build and integrate the GIS Project. After a series of negotiations, WWE and Centaura agreed to a cost-plus contract type with a cap of $7.6 million. Centaura was contracted to assist with collecting the business requirements; designing and building the system; designing and validating the business processes; developing and testing the system; creating training materials; deploying the system; and providing up to a 6-month warranty to fix any defects. WWE's primary responsibilities included providing the requirements; conducting training; and performing organization development activities to improve adoption, establish the supporting organization (with the help of Centaura), and implement the business processes to support the GIS.

Discussion Questions

1. When should a company build versus buy a product or service?
2. Do you agree with the decision to build a new system internally? Why or why not?
3. For the GIS Project, should the project team conduct a thorough due diligence before hiring Centaura? Why or why not?
4. Is the relationship with Centaura transactional or strategic? Explain.

Additional/Optional Questions of Considerations

5. The contract type of cost-plus was controversial. Ms. Holmes, for example, wanted a fixed cost contract, but Centaura disagreed citing the previous failure. Mr. Zhang eventually made the executive decision and agreed to the cost-plus contract. What do you see as the pros and cons of cost-plus?
6. Since the contract type is cost-plus, is it important to implement a robust change management plan? Why or why not?
7. Given the past issue with the ERP vendor, what will you do differently with Centaura?
8. Since Centaura is performing the bulk of the development work, how do you, as the project manager, manage the vendor's work and to make sure the performance report reflects the actual situation?

Exercise

1. Comprehensively identify the procurement issues and risks for this project knowing the cost-plus contract type will be utilized.

XVIII. Leveraging Conflicts

Early in the project, there was clear tension among the project team members. Some individuals were both elated to have a second chance at tackling this project, but they also remembered the stormy days of the past. Perhaps out of an abundance of caution or perhaps ill will as a result of the earlier failure, the project team members disagreed over a range of areas.

Discussion Questions

1. Using the Thamhain and Wilemon's seven conflict types as shown in Table 18.1 Type of Project Conflicts by Thamhain and Wilemon on page 225 and given the information about this project, identify and describe one possible conflict for each of the seven conflict types.
2. Map the seven conflicts type above to Jehn's three conflict types: task, relationship, and process. Describe which topology do you prefer and explain why.
3. Evaluate the most complex conflict that you have identified above and describe the potential interaction between the multiple types of conflicts.
4. Discuss how you plan to tackle this conflict. Which approach would you adopt? Explain why.

Additional/Optional Questions of Considerations

5. As the project manager, how can you leverage the history of this project to your advantage? Are there positive aspects of conflicts on this project?
6. Are there conflict situations in which avoiding is the best approach?
7. Are there conflict situations in which forcing is the best approach?
8. On this GIS project, is there a situation where you, the project manager, instigates and encourages conflicts? Explain.

XIX. Governance Management

The GIS project is a project within the larger ERP project, and consequently, there are potentially multiple layers of project governance.

Discussion Questions

1. Why is governance important for this project?
2. Discuss three good principles that are must applicable to this project.
3. Based on your understanding of this project and the organization, how many layers of governance do you propose for this project? Explain.
4. Who (name or roles) should participate in the layer(s) of governance that you proposed earlier?

Additional/Optional Questions of Considerations

5. When should you, as the project manager, formally establish the governance for this project?
6. What is the role of a project manager in the governance team?
7. What kind of project situations or decisions require governance team approval?
8. Given the interest in this project, there are likely to be more involvement than less. As the project manager, how do you balance the multiple and potentially conflicting interests?

Exercise

Develop a project governance plan for this project. See Template 12: Project Governance Plan on page 281 in Appendix A.

XX. Working with People

Even though the GIS Project is a combination of software development and business process design, Ms. Holmes has warned that this project should be viewed through the lens of organization development because the GIS data types will be widely used and their adoption will be crucial for its success. With Mr. Zhang's support, you are selected to lead the project.

As you are performing the detailed project planning, you are also evaluating how to best lead the project teams and engage the stakeholders around the world. So far, while there have been some minor issues and complaints, nearly everyone has thought you have performed admirably in the preparation phase by being transparent and inclusive. But you know that once implementation starts, the various problems that you have already experienced will flare up.

Discussion: As you are the project manager…

1. Since this project has high visibility and involves many stakeholders, discuss three activities each that highlights the distinction between project leadership and project management.

2. Explain the leadership style that you plan to use in the execution phase.
3. Describe how you plan to integrate the vendor project team (Centaura) with WWE team members.
4. One of the lessons from the previous ERP project was a lack of clarity on the specific roles and responsibilities. How do you plan to address this concern for this project?

Additional/Optional Questions of Considerations

5. Identify a likely people issue that will come up in this project. Suggest an optimal way of addressing it.
6. How would you effectively conduct a project kick-off meeting?
7. What tactics would you employ to build a cohesive project team? How about a high-performance team?
8. With many executives, including the CPO, paying close attention to this project, what is your plan of effectively "managing up"? Explain.

Exercise

Prepare a project kick-off presentation to jump start the project.

XXI. Beyond Project

Ms. Holmes was reviewing the lessons learned, and a number of suggestions caught her attention. These suggestions and comments can be grouped into two groups: operation management and program management.

- On operation management, there were some thoughtful comments about the start of the planning of the operation much earlier in the project life cycle. In fact, one suggestion even proposed to plan the project with the business operation as the starting point. Even though the overall deployment was much better than the failed ERP, there were still some significant issues. Luckily, only a handful of those incidents became visible; the project and the operation team worked a minimum of 60 hours a week per person for an entire month to address the incidents. While not quite a "death march", some team member did use the term to describe the intense situation. With GIS Project Phase 2 coming, this is an area of concern. Even without reading through the lessons learned, Ms. Holmes was already thinking of applying strong service management processes in the information technology departments.
- On program management, even though, in this case, the GIS is positioned as a subproject within the ERP project, the better arrangement is a program management structure. Specifically, GIS is one project within the overall

ERP program, and there can be other projects. Here the comments on how GIS is managed as a program were much more subtle. There were universal praises on the effectiveness of project management on GIS. But reading at a more thoughtful level, there were a significant number of integration, governance, and strategic alignment concerns, especially with some existing systems. For example, the prioritization of some project integration between GIS and other systems was largely based on "who shouted the loudest" and not necessarily on business priorities. Worse, some major activities were not coordinated very effectively. Specifically, the training team completed the bulk of the training development before the business requirements were signed off. This led to significant rework of the training materials.

Of course, given the previous failed ERP project, the expectation for GIS Phase 1 Project was relatively low. GIS Phase 1 was successful, but there were too many concerns lurking right below the surface. Now with its successful launch, expectations will naturally rise. Thus, Ms. Holmes inherently knows that she must improve the project execution further in future projects.

Discussion Questions

1. Should Ms. Holmes be concerned about these project concerns that were lurking below the surface? Or should she attribute them to continuous improvement?
2. On large and intense projects, how can project managers minimize the likelihood of prolong days for a long stretch of time?
3. When should a project plan for its operation, also known as the "the day after" project launch?
4. What are the similarities and differences between project management and program management?

Additional/Optional Questions of Considerations

5. Should the organization adopt some processes in Information Technology Infrastructure Library (ITIL)? If yes, highlight three of them.
6. Describe what "execution excellence" mean for this project?
7. If you are in Ms. Holmes position and preparing for the GIS Phase 2 Project, will you continue to use project management or will you use program management?
8. Should Ms. Holmes consider creating an enterprise Project Management Office as she prepares to tackle the next phase of GIS as well as other large projects in WWE?

Appendix C: Glossary of Key Terms

The term defined here are reprinted from *The Sensible Guide to Key Terminologies in Project Management*, with permission from the authors.

Term	*Description*
Accuracy	An assessment of correctness used in estimation and quality management systems.
Activity	A basic unit of work in project planning that serves as the basis for scheduling, resource planning, and budgeting. An activity usually comprises numerous related tasks, which must be completed in order for the activity to be considered as complete.
Agile	A management philosophy that focuses on value and customer interactions where requirements evolve through the collaborative effort of self-organizing cross-functional teams. The term was popularized in the Agile Manifesto.
Agile Certified Practitioner (PMI-ACP®)	Project Management Institute's professional-level certification for project professionals who specialize in Agile approaches and methods. This certification is designed to indicate the individual's ability to lead Agile projects including the application of Agile management concepts, processes, tools, and techniques. Unlike many other Agile-related certifications, PMI-ACP is methodology neutral.

(*Continued*)

Term	*Description*
Agile Manifesto	An original proclamation that articulates four key values and twelve principles that software developers should use to guide their work.
Analogous estimating	A technique for the estimation of a range of project measures largely based on historical data of similar activities or projects. The project measures can include cost, duration, resources, and scope.
Assignment	A defined unit of work that has the following attributes: start and end dates, clear objectives and outcomes, people or roles assigned to work on it, and associated performance metrics.
Baseline	An approved set of project work to be tracked and compared with the actual results. Any change to the baseline requires a formal change management authorization.
Benefit	The tangible and intangible gains such as financial advantages, new products, competitive capabilities, and valuable additions received by the stakeholders and the organization as a result of the program.
Bottom-up estimating	An estimating technique whereby the work breakdown structure's lower-level tasks and activities are aggregated in order to estimate the cost or duration of a project. This technique is generally more accurate than analogous and parametric estimating.
Brainstorming	A group creativity technique used to gather information, often for solving problems by identifying threats, determining root causes, and generating options and solutions with the help of subject matter experts and specialists.
Budget	The approved estimate for the total project, program, or portfolio costs that is inclusive of the cost of all planned components. It also includes the total amount of component cost estimates, contingency reserves, and sometimes the management reserve as well.
Capacity and capability analysis (CCA)	An analysis technique designed to evaluate the resources (human, finance, technology, machinery) of an organization required to implement projects, programs, or portfolios.

(*Continued*)

Term	*Description*
Certified Associate in Project Management (CAPM®)	Project Management Institute's entry-level certification for project professionals who command the basic knowledge of project management concepts including knowledge areas, processes, tools, and techniques.
Change control	The process of managing project change, including key documents, deliverables, and baselines associated with the initiative. The process includes the identification, documentation, change assessment, and decision-making.
Change request	A proposition to modify a previously agreed-upon baseline, document, or deliverable.
Conflict	Escalating disagreements arising from differences in priorities, processes, and personal or organizational views and values. If left alone, conflicts can spiral into larger conflicts with vicious negative cycles feeding and intensifying themselves.
Constraint	A factor that limits the options or establishes the boundaries for managing a project, program, portfolio, activity, or process.
Contingency	An uncertain event or occurrence that may affect the project, program, and portfolio execution in which the project professional should consider setting aside a reserve should the event occur.
Contingency plan	A preconceived plan that describes ensuing actions should some predetermined trigger conditions occur.
Contingency reserve	An active risk response strategy in which time or funds are proactively set aside to manage the known project, program, or portfolio risks.
Cost of quality (COQ)	The cost associated with managing and delivering a quality product or service that includes quality planning, quality control, and quality assurance.
Cost Performance Index (CPI)	Earned value management measure designed to evaluate project or program cost efficiency. It is expressed as a ratio of earned value to actual cost.
Crashing	A scheme for schedule compression intended to decrease the total period of time for the minimum incremental cost by adding additional resources to the task or activity.

(Continued)

Term	*Description*
Critical path	The longest activity sequence of a project or program that determines the shortest possible duration to complete the endeavor.
Culture	The system of values, behaviors, attitudes, and traditions that are often unspoken in an organization, which affects the planning and execution of projects, programs, and portfolios.
Decomposition	A planning scheme that deconstructs abstract work into smaller and more concrete components.
Definitive estimating	The cost estimate build from the detailed level analysis of tasks and activities, typically from a work breakdown structure of a project and program. Definitive estimates are the most accurate methods of estimation with a range of –5% to +10% of the predicted cost.
Deliverable	Any distinct and unique element, product, item, or result that is developed for delivery at the completion of a particular project activity, component, task, or at the completion of the whole project.
Duration (DU or DUR)	The total amount of time required to complete work such as a project, an activity, a task, an assignment, or a work breakdown structure component. Duration is expressed in a unit of time (e.g., years, months, weeks, days, hours, or minutes).
Earned value management (EVM)	The project management performance methodology that provides integrated management of scope, schedule, and resource measurements to evaluate a project and program performance and progress.
Estimate	A quantitative assessment of approximate project, program, or portfolio costs, resources, or duration.
Fit for purpose	Suitability of a product for a particular use. For example, the ability of the customer to use the software for a particular purpose.
Fit for use	Suitability of a product, in its present condition, for a particular use. For example, the ability of multiple customers to use the software simultaneously without significant performance deterioration.

(*Continued*)

Term	*Description*
Float	The duration of time for which a delay can be incurred by a task in the project network without delaying the completion date as well as subsequent tasks of the project.
Governance	The alignment of project, program, or portfolio goals with the sponsoring organization's strategy through sound decision-making processes on authorization, oversight, resource allocation, and change management.
Ground rules	Basic rules and expectations for acceptable team behavior.
Hybrid approach	A framework based on multiple Agile and non-Agile components, typically resulting in a non-Agile output that manifests some benefits of applying Agile.
Issue	An occurring or occurred incident of a project or program that if unmanaged will affect the schedule, scope, cost, resources, or other project parameters.
Iteration	A developmental cycle of a product in which all required activities are carried out within a prespecified block of time.
Key Performance Indicator (KPI)	A key measure as to how effective the performance of a project is in regard to agreed-upon, required, and identified strategic objectives. KPIs are often confused with critical success factors (CSFs). Think of CSFs as "causes" for success, while KPIs are indicators that measure the "outputs" of striving toward that success.
Lessons learned	Lessons learned is both an activity and a technique whereby a project, program, or portfolio is critically evaluated with the intent of understanding what worked well and identifying areas for future improvement. Prior to the start of a future initiative, a leading practice is to review the lessons learned knowledge base.

(*Continued*)

Term	*Description*
Life cycle	The process through which projects, programs, portfolios, and products are conceived, designed, implemented, and eventually terminated, from a beginning to an end. The life cycle is temporary by definition, even though some endeavors have significantly longer durations. A representative and generic project, program, and portfolio life cycle, which can include these five phases: Ideation, Initiation, Preparation, Implementation, and Transition. Transition can vary depending on the nature of the project, program, and portion. Three common options include closure, transition (to another project, program, portfolio, or operation), and optimized (when project, program, and portfolio merge with another endeavor).
Management reserve	An active risk response strategy in which time or resources are proactively set aside to manage the unplanned or unknown project, program, or portfolio risks. Management reserves are generally held by the next-level management and are financial in nature but can extend to human, technical, and equipment resources.
Milestone	A significant moment in time of a project, program, or portfolio, such as the completion of a major activity or phase.
Operation	The business function of managing the regular, often cyclical and routine activities of an organization.
Opportunity	A potential situation or condition that is favorable to one or multiple project objectives.
Organizational project management (OPM)	A structure in which project, program, and portfolio management are incorporated with organizational enablers for accomplishing strategic goals and objectives.
Parametric estimating	An estimation scheme involving an algorithm for calculating duration or cost on the basis of project or program parameters and historical data.

(Continued)

Term	*Description*
Parkinson's law	A general law that describes a phenomenon in which work somehow expands to fill the allotted amount of time.
Plan–Do–Check–Act (PDCA)	A continuous improvement framework, which provides a simple and effective approach for managing change and solving problems. The framework can also be applied to iterative product development, to incremental project management, and for testing improvement measures on a small scale before deploying to a broader context.
Portfolio	A logical collection of programs, projects, operations, subsidiary portfolios, and other related work that should be managed collectively to achieve one or more of organizational and strategic goals and objectives.
Portfolio management (PfM)	The centralized streamlining of one or multiple portfolios for the achievement of strategic goals and objectives by incorporating the concepts of identifying, prioritizing, authorizing, monitoring, and controlling of the entirety of the portfolio.
Portfolio Management Professional (PfMP®)	Project Management Institute's professional-level certification for portfolio managers. This certification is currently the apex of the project management professional ladder, especially with an emphasis on business and governance processes. Attaining this certification is a demonstration of leadership, organizational savviness, and business acumen to make business investment decisions and lead their implementation.
Precision	The measure of exactness. Measurements that are close to each other are said to be precise.
Product backlog	Commonly used in Agile, it contains a listing of all activities, tasks, new features, changes to existing features, or bug fixes, which need to be completed to fulfill project or product requirements satisfactorily or to achieve a specific outcome.
Professional in Business Analysis (PMI-PBA®) (PMI-PBA®)	Project Management Institute's professional-level certification for project business analysts who specialize in the development of project and product scope and requirements.

(Continued)

Term	*Description*
Program	A collection of highly related components, such as projects, subprograms, and other activities. When these components are managed as a program, they can achieve great value and benefits not possible if they were managed separately.
Program Evaluation and Review Technique (PERT)	A method in which a weighted average of pessimistic, optimistic, and most likely activity duration is taken into consideration for the estimation of project duration when the individual estimates involve uncertainty.
Program management (PgM)	The coordinated management of program activities by applying specific principles, knowledge, processes, tools, and skills to deliver results effectively.
Program Management Professional (PgMP®)	Project Management Institute's professional-level certification for program managers. Achieving this certification indicates the professional's competency at leading a group of related projects and activities and to achieve greater value than managing these components individually.
Project	A time-limited, purpose-driven, and often unique endeavor intended to create an outcome, service, product, or deliverable.
Project charter	A document issued by an individual or group responsible for sponsoring or initiating the project. The project charter grants formal authorization to the project manager to guide and oversee activities within the context of organizational, contractual, and third-party resources.
Project management	A management discipline specialized for overseeing projects. Project management includes the activities associated with initiating, planning, executing, monitoring, controlling, and closing the work for meeting the project goals and objectives by applying skills, knowledge, techniques, and tools to project activities and tasks.

(*Continued*)

Term	*Description*
Project management office (PMO)	A management framework that assists in the sharing of methodologies, leading practices, resources, techniques, and tools as well as the standardization of the governance processes that are related to the project. Note: The general acronym PMO can represent project-, program-, and portfolio Management Office. This dictionary makes the distinction and correspondingly, the acronym used is PMO, PgMO, and PfMO (project, program, and portfolio respectively.)
Project Management Professional (PMP®)	Project Management Institute's professional-level certification for project managers. This certification is designed to cement a working knowledge of project management and the leadership competencies to direct projects.
Quality	The extent to which a set of attributes produced by projects and programs meet the stakeholder, business, functional, nonfunctional, and transition requirements.
RACI	A specific type of responsibility assignment matrix (RAM) that is also an acronym where R stands for Responsibility, A for Accountability, C for Consulted, and I for Informed.
Request for Information (RFI)	A procurement process and the associated document whereby the buyer requests various information on a product or service from potential sellers, typically without a commitment to purchase. In most situations, the Request for Information (RFI) occurs very early in the procurement process where the buyer and seller may even meet for the first time.

(*Continued*)

Term	*Description*
Request for Proposal (RFP)	A procurement process and the associated document whereby the buyer requests specific information and pricing on a product or service from prospective sellers. Typically, the Request for Proposal (RFP) occurs later in the procurement process where mutual interests are already determined, often following the Request for Information (RFI) and Request for Quotation (RFQ) process. In some instances, this process and document can be formal.
Request for Quotation (RFQ)	A procurement process and the associated document whereby the buyer requests specific indicative or approximate pricing on a product or service from prospective sellers. Typically, the Request for Quote (RFQ) follows the Request for Information (RFI) process.
Requirement	A condition or capability that is required to be present in a product, service, or result to satisfy a contract or other formally imposed requirement. Commonly categorized as stakeholder, business, functional, nonfunctional, and transition.
Reserve	Financial and nonfinancial resources held by the Governance team to address the project, program, and portfolio risks.
Resource leveling	A technique for resource optimization to adjust hours available for work per team member, and the number of hours allocated to each individual, which may influence the critical path.
Responsibility Assignment Matrix (RAM)	A table that describes a project or program resource assignments on tasks, activities, deliverables, or work packages.
Risk	A potentiality that, if it materializes, can have an impact on one or multiple objectives in a negative or positive manner, in the form of resources, performance, quality, or timeline.

(*Continued*)

Term	*Description*
Risk management	The practice of identifying, evaluating, prioritizing, and monitoring unknown and probable events that may affect projects, programs, and portfolios. Once prioritized, selective risks are analyzed further to develop a risk response plan and when necessary implement them to mitigate threats and exploit opportunities.
Risk register	A repository that captures the essential details of the project, program, and portfolio risks.
Rolling wave planning	An iterative technique that involves progressive elaboration for adding details to the work that has to be completed in the near-term on an ongoing basis, while a higher level for planning is used for work that has to be done in the future.
Schedule compression	An approach to project scheduling, whereby the duration is shortened, while not minimizing and/or reducing the scope or quality of the project in any way.
Schedule Performance Index (SPI)	Earned value management measures designed to evaluate project or program schedule efficiency. It is expressed as a ratio of earned value to planned value.
Scheduling Professional (PMI-SP®)	Project Management Institute's professional-level certification for project scheduler who specializes in project scheduling methods and the ability to manage time. This certification is designed to indicate the individual's ability to manage schedules of large and often complex projects.
Sponsoring organization	A particular group or enterprise that is explicitly and directly involved in the funding and execution of the project, program, or portfolio. This term is also referred to as the "performing organization".
Strategic alignment	The process of linking and evaluating organizational strategy with project, program, and portfolio goals and objectives, taking into account the business environment and any implementation challenges.

(*Continued*)

Term	*Description*
Strategic business execution (SBE)	An interdisciplinary framework to deliver consistent and sustainable business results through focus and prioritization, close alignment between strategy and implementation, thoughtful planning, disciplined execution, and sensible agility to manage changes and unknowns.
Technique	A defined procedure to perform a particular task and activity, such as the execution or performance of an artistic work or a scientific procedure and often by using specific tools.
Threat	A risk that has the potential to influence negatively one or multiple portfolio, program, or project goals.
Timebox	A technique for fixing time in which a task or activity must be accomplished.
Tool	A tangible aid, such as a template or software, used in performing a task or activity to more effectively and efficiently produce a product or a result.
User story	An informal, natural language description of one or more features of a software system. User stories are designed to provide clarity and real-life use of how the user will interact with the system.
Value	Tangible, intangible, and sometimes abstract business gains and advantages that are associated with the sponsoring organization's strategy, goals, and objectives.
Vision	A description of the direction in which an organization is intended to move toward (e.g., "Where do we want to be?").
WBS dictionary	A detailed document that supports the work breakdown structure (WBS) by enlisting the details of the deliverable, activity, and scheduling information about WBS components.
Work package	The lowest segment of the work breakdown structure (WBS) for which the estimation and management of duration and cost are carried out.

Index

D

E

F

G

H

I

T